Lukas Grünhaupt

Granular aluminium superinductors

Experimental Condensed Matter Physics
Band 26

Herausgeber
Physikalisches Institut
Prof. Dr. David Hunger
Prof. Dr. Alexey Ustinov
Prof. Dr. Georg Weiß
Prof. Dr. Wolfgang Wernsdorfer
Prof. Dr. Wulf Wulfhekel

Eine Übersicht aller bisher in dieser Schriftenreihe erschienenen Bände finden Sie am Ende des Buchs.

Granular aluminium superinductors

by

Lukas Grünhaupt

KIT Scientific Publishing

Karlsruher Institut für Technologie
Physikalisches Institut

Granular aluminium superinductors

Zur Erlangung des akademischen Grades eines Doktors der Naturwissenschaften von der KIT-Fakultät für Physik des Karlsruher Instituts für Technologie (KIT) genehmigte Dissertation

von M. Sc. Lukas Grünhaupt

Tag der mündlichen Prüfung: 14. Juni 2019
Referent: Prof. Dr. Alexey V. Ustinov
Korreferent: Prof. Dr. Benjamin Huard
Betreuer: Dr. Ioan M. Pop

Impressum

KIT Scientific Publishing

Karlsruher Institut für Technologie (KIT)
KIT Scientific Publishing
Straße am Forum 2
D-76131 Karlsruhe

KIT Scientific Publishing is a registered trademark of Karlsruhe Institute of Technology.
Reprint using the book cover is not allowed.

www.ksp.kit.edu

Print on Demand 2019 – Gedruckt auf FSC-zertifiziertem Papier

ISSN 2191-9925
ISBN 978-3-7315-0956-1
DOI 10.5445/KSP/1000097320

Abstract

Superconducting quantum circuits operated at microwave frequencies are currently one of the most promising approaches towards the realization of quantum information processing devices. A potentially limiting factor of prevalent circuits with low characteristic impedance $\sim 100\,\Omega$ is their small anharmonicity, which might limit their coherence and logic gate fidelity. Recently, circuits with high characteristic impedance, larger than the resistance quantum $R_Q = h/(4e^2)$ begun to attract interest as a possible alternative. Besides the basic building blocks of superconducting circuits, which are capacitors, inductors, and Josephson junctions, high impedance circuits also include so-called superinductors, which feature an inductance $\sim 10^2\,$nH, and characteristic impedance $Z > R_Q$. Superinductors are mostly realized using compact arrays of large Josephson junctions, and remarkable results were achieved, albeit their complexity in design and fabrication.

Here, we investigate superconducting granular aluminum (grAl) as a promising material for the implementation of superinductors. Granular aluminum, i.e. pure aluminum grains in an insulating AlOx matrix forms by depositing pure aluminum in an oxygen atmosphere. Due to the amenable kinetic inductance of grAl wires, they are a straightforward technique for the realization of superinductors. Using microwave resonators, we show that grAl strips can reach kinetic inductances up to nH/□, while their microwave frequency losses are as low as state of the art superinductor implementations. We identify excess quasiparticles as a limiting loss mechanism in superconducting circuits employing grAl, and find quasiparticle relaxation times on the order of seconds, orders of magnitude longer than previously observed.

Furthermore, to test a grAl superinductor we use it in a superconducting fluxonium quantum bit, which consists of a small Josephson junction shunted by a superinductor. The measured coherence times of our grAl fluxonium are comparable to fluxonium implementations using Josephson junction array superinductors, and demonstrate that granular aluminum is a viable material for the implementation of superinductors. We believe that granular aluminum superinductors will enable the realization of increasingly complex, and potentially protected quantum bit designs, which possibly present a new approach towards scalable quantum information processing circuits.

Acknowledgements

Looking back today I consider myself very lucky. Lucky because of the people I worked with, learned from, and was supported by. Even though a PhD project sometimes feels like a solitary journey, in my opinion it is by all means a team effort.

First of all thank you to my advisor Ioan Pop. During a visit to Ioan's previous group shortly before the start of my PhD all of his former students congratulated me on the chance to work with him. Today I know why, and can only repeat the statement I most often heard: "He is a great guy to work with, and you are going to learn a lot!". Thank you for all your time, patience, knowledge, and encouragement over the last three years. From the "early" days, when he taught me how to wire a dilution fridge or how to make the clean room fabrication work, to later when we discussed measurement results, I always felt I gained some insight or new perspective. I can't remember a time when he wasn't available for a quick question or to look at some data together. Although I will forever remember his favorite phrase: "Let's just quickly measure/plot/calculate this", I will also forever remember that most of the time when he dropped the phrase it was him who was sitting next to me in the lab at 11:00 PM trying to finish something. I can only begin to describe how much I learned during my time as Ioan's PhD student. Thank you for being an excellent teacher and advisor, and always leading by example when it comes to working hard.

I would like to thank Prof. Alexey Ustinov not only for being my first reviewer, but also for accepting me as a master student before I became a PhD student. I learned a lot from being a member in his group, and from all the seminars I attended over the last years. His profound knowledge has sparked many interesting discussions, and added new perspectives to ongoing projects. Thank you also to Prof. Benjamin Huard, who accepted to be my second reviewer. I am very grateful for the considerable effort and time he invested for my PhD defense.

Like many PhD students before me I would like to thank the administrative staff of the Physikalische Institute. Especially, I would like to thank Steffi Baatz for all her help in ordering various things from screws to VNAs, in figuring out how to get some paperwork, and for dealing with the higher layers of KIT bureaucracy. A huge thanks to Hans-Willi Pensl and Christoph Sürgers for taking such good care of our old helium liquefier and repairing it time after time. I would also like to thank Roland Jehle for help with all the electronics and ordering numerous small pieces. I am grateful for Michael Meyer running our mechanical workshop, his support in the design of anything from sample holders to cryostat parts, and squeezing the occasional super urgent part into the order book.

Thank you to all past and present fluxons. I feel very fortunate to have worked with such a great group of people. Discussions about physics certainly dominated, but the lunch and coffee breaks were often spent venturing into all sorts of entertaining discussions. When we were outside the lab at conferences or extracurricular activities, the times were always memorable. I think it's fair to say we were not only colleagues, contrary to what I once allegedly said. A special thanks goes to Hannes Rotzinger. Before I ever knew what kinetic inductance was, he had started research about granular aluminum in Karlsruhe, which I certainly benefited a great deal from. Aside from that, I am thankful for the numerous practical tips, discussions, and occasional chats and words of encouragement. I am sure the 4th floor wouldn't be the same without him there. Thank you also to Sasha Lukashenko, the cryostat whisperer, as I like to call him. He had set up our Magpie fridge perfectly, taught me how to operate it, and was there for assistance when needed. His extraordinary micromechanical skills brought us all sorts of filters small enough to fit in our cryostat. Finally, thank you to (now) Prof. Martin Weides, who was my advisor during my Master thesis, and allowed me to gain my first scientific working experience. The basic skills I learned during my time as a master student proved instrumental for my PhD project.

I want to thank all Black Forest quantum members for their camaraderie and help. Thank you Dasha, Francesco, Ivan, Fábio, Dennis, and Julian. Especially, I want to thank Natasha Maleeva for her words of encouragement, help with any detail, no matter how small, and profoundly positive attitude. The office always felt a little nicer with Natasha there, and I certainly enjoyed our discussions about physics and all the other aspects of life.
It's great to call Martin Spiecker a colleague now, after he worked with me during his Master thesis. I think our collaboration started out very successfully, and in no small part thanks to his clever ideas. Thank you to Patrick Winkel for his help in professional matters ranging from taking the most beautiful sample pictures, to producing truly user-friendly parametric amplifiers, or helping with any task in the lab. Having you there as a sparring partner to trade friendly insults and for the occasional ping-pong match made the working day so much more enjoyable. His detailed feedback on practice talks or manuscripts would often lead to hour-long discussions, which were certainly always appreciated.

Thank you to my Bruhrain friends back home. No matter how seldom I see them or how many occasions I miss, they always make me feel welcome. Thank you to Michael Jenne, a constant source of good vibes from the early days of elementary mechanics in our first semester to memorable road trips along the way. I couldn't have asked for a better companion. Thank you to all my friends and teammates of Engineers football. Having to go to practice more than once kept me from obsessing about all the small and big problems you encounter during a PhD.

Mira, thank you for your support over the last years. I could not ask for a better partner. You provided support and encouragement in hard times, but also made sure I would not get completely absorbed in problems. By accepting all my other activities on top of my long working hours you gave me the freedom to push for my goals but never made me feel like I was alone in any of it. Seeing you work hard and focused towards your goals is something that inspires me. With your trust and sense of adventure, I am sure whatever journey we embark on next will be a great one. I am more than grateful to have you by my side.

In the end, where would I be without my family. Thank you to my sister for being a much better little sister than I was a big brother. Thank you Mama and Papa for your unwavering support. You provided a safe haven to start all sorts of adventures from, made me who I am today, and never once doubted me along the way. Thank you for believing in me, when I didn't. I am only where I am today thanks to your love and support.

Karlsruhe, June 2019 Lukas Grünhaupt

Contents

Abstract . . . . **i**

Acknowledgements . . . . **iii**

1 Introduction **1**

1.1 Superconducting qubits 4

1.2 The fluxonium qubit 6

1.2.1 Quantum fluctuations of charge and phase 6

1.2.2 The uncoupled fluxonium Hamiltonian 7

1.2.3 Fluxonium qubit loss channels 12

1.2.4 Inductively coupled fluxonium-resonator system 15

1.3 Granular aluminum for high impedance environments 18

1.3.1 Status quo of high impedance environment realization 18

1.3.2 Kinetic inductance and nonlinearity of granular aluminum 24

1.4 Loss mechanisms in granular aluminum 28

1.4.1 Resonator measurements to evaluate losses 28

1.4.2 Dielectric loss 31

1.4.3 Excess quasiparticles in grAl resonators 34

1.5 Fluxonium qubit case study 37

1.5.1 Design 37

1.5.2 Spectroscopy 39

1.5.3 Coherence 40

2 Granular aluminum in the literature **45**

2.1 Overview of experimental results 45

2.2 Possible origins of enhanced critical temperature 49

2.3 Suppression of T_c in high resistivity granular aluminum 51

3 High impedance granular aluminum resonators **53**

3.1 Design 53

3.1.1 Sample holder 53

3.1.2 FEM simulation of designs employing kinetic inductance 55

3.1.3 Rectangular strip resonators . . . 57
3.1.4 Microwave kinetic inductance detectors . . . 58
3.1.5 Coplanar waveguide resonators . . . 60
3.2 Measurement results . . . 60
3.2.1 Rectangular strip resonators . . . 60
3.2.2 MKID geometry resonators . . . 70
3.3 Dielectric loss in grAl resonators . . . 75
3.4 Quasiparticle dynamics in grAl resonators . . . 78

4 Granular aluminum fluxonium qubit . . . 87
4.1 Design and fabrication . . . 87
4.1.1 Readout resonator . . . 88
4.1.2 Granular aluminum fluxonium . . . 90
4.1.3 Superinductor implementation techniques . . . 91
4.2 Spectroscopy . . . 97
4.2.1 Sample 1 . . . 97
4.2.2 Sample 2 . . . 100
4.3 Coherence . . . 101
4.3.1 Sample 1 . . . 101
4.3.2 Sample 2 . . . 106

5 Conclusion & Outlook . . . 111

Appendix . . . 115
A Fabrication techniques . . . 115
A.1 Granular aluminum deposition . . . 115
A.2 Resist residue cleaning . . . 116
A.3 Electron-beam lithography lift-off . . . 117
B Phenomenological model of quasiparticle activation . . . 120
B.1 Steady-state . . . 121
B.2 Decay rate . . . 122
B.3 Comparing to experiments . . . 122
C Experimental setup . . . 124
C.1 Cryogenic microwave wiring . . . 124
C.2 Timedomain setup . . . 128

Bibliography . . . 129

List of publications . . . 145

1 Introduction

The ever increasing computational power of classical computers is transforming economic systems and societies around the globe, and has resulted in the advent of the *information age* at the beginning of the 21st century. Fueled by Moore's law, the digital revolution has lead to big data methods, and companies whose business model solely relies on the evaluation of immense amounts of data enabling increasing automation capabilities and efficiency.

However, as transistors get smaller and approach the limits imposed by the fundamental laws of quantum physics, the search for the next groundbreaking computational development continues. Quantum information processing, ultimately aiming for a fully *digital quantum computer* is currently one of the most exciting concepts. Due to the huge Hilbert space of 2^N dimensions spanned by N quantum bits (qubits), quantum computation promises to potentially enable the calculation of molecule synthesis, leading to new methods for drug design and potentially new technological avenues for the nitrogen fixation problem. Common to these applications is that they exceed the capabilities in terms of memory and computational speed even of today's most powerful supercomputers. The current threshold for brute force simulating of molecules is estimated at a number of atoms on the order of 50.

As Feynman said in 1982: '[...] *nature isn't classical, dammit, and if you want to make a simulation of nature, you'd better make it quantum mechanical,* [...]'. Fortunately, or unfortunately, depending on whether one agrees with Feynman when it comes to scientific problems: '[...] *by golly it's a wonderful problem, because it doesn't look so easy*' [1]. Although quantum computation has attracted significant attention and even reached the mainstream media, with some technology consulting firms predicting that actual applications and products will be reached in 5 to 10 years [a], there is not yet a clear path towards the actual physical implementation of a fully fledged quantum computer. Regardless, global companies like Google, IBM, Intel, and Microsoft have begun to invest in quantum computing, with start-ups and spin-offs also attracting substantial investments. One of the most promising approaches towards quantum computing today are superconducting circuits, which is also the technological avenue pursued by Intel, IBM and Google. However, after the remarkable

[a] http://www.gartner.com/smarterwithgartner/5-trends-emerge-in-gartner-hype-cycle-for-emerging-technologies-2018/ - accessed 3rd May 2019.

progress during the last two decades [2], the increase in complexity brought about by the scaling to larger and larger numbers of interconnected qubits has slowed down progress.

Since the demonstration of the first superconducting quantum bit (qubit) by Nakamura et al. in 1999 [3], the fundamental building block of a future quantum processor, the field of quantum information processing with superconducting devices has achieved an increase in terms of the operation time of these systems by 4 to 5 orders of magnitude. The most important metric still for the evaluation of single qubits is their coherence time, representing how long the quantum superposition state is conserved. Looking in more detail, the real metric is the ratio between coherence time and the operation, or gate time of a qubit. As superconducting circuits have shown a number of remarkable proof of principle experiments and are transitioning into the realm of higher integration depth and increasing number of qubits, a major point of emphasis is the *gate fidelity*, indicating how precisely an actual operation carries out a theoretically desired qubit manipulation. With increasing number of qubits on quantum chips, and despite considerable experimental effort, the fidelity of two-qubit gates is still below the threshold needed for quantum error correction [4]. Problems such as unintentional crosstalk between on-chip components, signal leakage and fluctuations of qubit parameters [5–7], necessitating frequent recalibration of superconducting qubit processors, are among the most pressing issues when it comes to further advancing the field.

Among superconducting circuits used for quantum information processing, the current workhorse of the field is the so called *transmon* qubit [8] invented in 2007. The transmon, an acronym for transmission-line shunted plasma oscillation qubit is essentially a slightly anharmonic LC-circuit whose lowest two eigenstates span the computational basis. Thanks to the considerable experimental effort that went into material, design, and fabrication optimization, transmon qubits and especially their two-qubit gates are on the verge of what is needed for today's quantum error correction schemes. Contrary to classical bits implemented using transistors, the no-cloning theorem for qubits prevents a straightforward error correction process of just redundantly encoding the qubit state into multiple physical copies and obtaining the correct result by a simple majority vote. The error correcting approach predominantly featured in industrial groups, namely the *surface code* [9] relies on a multitude of parity measurements to detect whether an error occurred or not. Still, at the current level of two-qubit gate fidelity in transmons, the overhead of encoding a logical qubit in an error corrected fashion is on the order of $10^3 - 10^4$ [9]. In fact, a major milestone for commercial companies working towards quantum computing with superconducting circuits is to demonstrate a smaller version of the error correction scheme.

A recent study investigating the resources and expected speedup of quantum computation for constraint satisfaction problems, which both belong to the NP-complete complexity class for classical computation, finds that there exists a substantial quantum speedup of a factor of 10^3. But, in order to achieve this speedup more than 10^{12} physical qubits are required assuming state of the art qubit performance [10]. Considering that the largest number of qubits currently combined on a single chip is 72 [b], significant improvements in terms of gate error rate and error correction schemes are necessary to imagine actual commercial use of quantum computers.

Motivated by the immense overhead needed for quantum error correction in current superconducting qubits, the investigation of alternative concepts realized in superconducting circuits is a very active research field with approaches ranging from the use of 3D cavity resonators [11–13] to the pursuit of topologically protected circuits [14]. Another concept gaining increasing attention recently, are high-impedance qubits such as the fluxonium qubit [15], which was invented almost 10 years ago. One of the basic building blocks in many of these alternative approaches is a so-called *superinductor*, which has a characteristic impedance larger than the resistance quantum $Z > R_Q$, where $R_Q = h/(2e)^2 \approx 6.5\,\mathrm{k\Omega}$. This value should be compared to the vacuum impedance of $Z_V = 377\,\Omega$, which is the natural scale for *LC* circuits. Therefore, superinductors are not easy to obtain. Not only is a large inductance of several hundreds of nH necessary, while any stray capacitance has to be minimized, but also one has to keep all self resonant modes of the superinductor well above the operating frequency range of the quantum circuit. From a design point of view, a superinductor should ideally act as a fully linear lumped element inductor. Up to now, superinductors have been implemented using arrays of Josephson junctions. Impressive results of circuits involving these types of superinductors have already been shown [16–18] and they continue to be a valuable part of the superconducting quantum circuit toolbox.

In this thesis however, we investigate granular aluminum, an emerging material in the superconducting quantum circuits community thanks to its ability to provide a high kinetic inductance, which enables characteristic impedances larger than R_Q. Using granular aluminum for superinductors could simplify the design and fabrication process of these circuit elements while potentially also reducing the technical complexity to a simple zero angle metal film deposition. Additionally, granular aluminum provides a large kinetic inductance at a significantly reduced nonlinearity compared to Josephson junction superinductors. We characterize the microwave frequency loss mechanisms in granular aluminum, to assess if they could potentially limit the coherence of superconducting circuits employing this material.

This thesis is organized as follows: the introductory chapter summarizes the main concepts and results obtained during the course of this thesis. The following chapter provides a selective literature survey of granular aluminum, which has been investigated since the 1960's in different physical contexts. In Chapter 3 we discuss in more details measurements on

[b] http://ai.googleblog.com/2018/03/a-preview-of-bristlecone-googles-new.html - accessed 3rd May 2019.

superconducting resonators with high characteristic impedance and conclude the description of the experimental results in Chapter 4 with a discussion of our case study of a superconducting high-impedance qubit, employing granular aluminum for its superinductor: the fluxonium. Finally, we summarize the results and provide an outlook of possible future areas of investigation and development for superconducting circuits using granular aluminum.

1.1 Superconducting qubits

The basic building blocks of superconducting quantum circuits, enabling a variety of systems, are the following components: capacitors, inductors, and the Josephson junction [19], which for the purpose of superconducting qubits is essentially a dissipationless, coherent, nonlinear inductor. Combining these three building blocks in different configurations and with varying ratios between the elements characteristic energy scales allows for the realization of different flavors of superconducting qubits.

At the starting point of the field of superconducting qubits lies the so-called Cooper-pair box, which was experimentally realized in 1999 [3, 20]. A superconducting island is connected to ground through a Josephson junction, which allows dissipationless and coherent tunneling of Cooper-pairs on and off the island. Using the circuit quantization technique in Ref. [21] we can obtain the circuit Hamiltonian,

$$\hat{H} = \frac{1}{2C}\hat{Q}^2 - E_\mathrm{J}\cos 2\pi\hat{\Phi}/\Phi_0, \tag{1.1}$$

with C the total capacitance of the Cooper pair box island. The Josephson junction term is characterized by its energy scale $E_\mathrm{J} = (\hbar I_\mathrm{c})/2e$, with I_c the critical current of the junction. Furthermore, the charge $\hat{Q}$ and flux $\hat{\Phi}$ are conjugate variables and obey the relation $[\hat{\Phi}, \hat{Q}] = i\hbar$. The generalized magnetic flux across a Josephson junction, given by the integral of the voltage, is associated with a phase φ across the element. Both variables are connected by the relation $\hat{\varphi} = 2\pi\hat{\Phi}/\Phi_0$ [21]. By adding a gate electrode to the circuit, an additional tuning parameter is introduced, namely the gate charge Q_g, which changes the charge term in the Hamiltonian to $(\hat{Q} - Q_\mathrm{g})^2/(2C)$.

A natural basis for this Hamiltonian is the Cooper pair number basis, where the energy associated with the charge is diagonal, while the Josephson junction introduces off-diagonal coupling terms. Introducing $\hat{N} = \hat{Q}/2e$, the Hamiltonian has the following form,

$$\hat{H} = 4E_\mathrm{C}(\hat{N} - n_\mathrm{g})^2 - \frac{E_\mathrm{J}}{2}\sum_N |N\rangle\langle N+1| + |N+1\rangle\langle N|, \tag{1.2}$$

Figure 1.1 shows the electrical schematic of the Cooper-pair box qubit and the energy level diagram as a function of the gate charge $n_g = Q_g/(2e)$. For small E_J/E_C we observe two effects, first at $n_g = \pm 0.5$ a gap opens up between two neighboring parabolae (grey). The transition energy between these two states is significantly different compared to the higher transitions, expressed as the systems anharmonicity $a = E_{01} - E_{12}$. For circuits with large a, the two lowest levels form an effective two-level system, well separated in frequency from the higher levels, allowing the use of this system as a qubit. Secondly, Fig. 1.1 shows a rapid change in the transition energy as a function of the gate charge. Since any fluctuation in the qubit transition frequency leads to decoherence, and charge fluctuations are ubiquitous in nature, the Cooper-pair box in the regime of small E_J/E_C is challenging to operate.

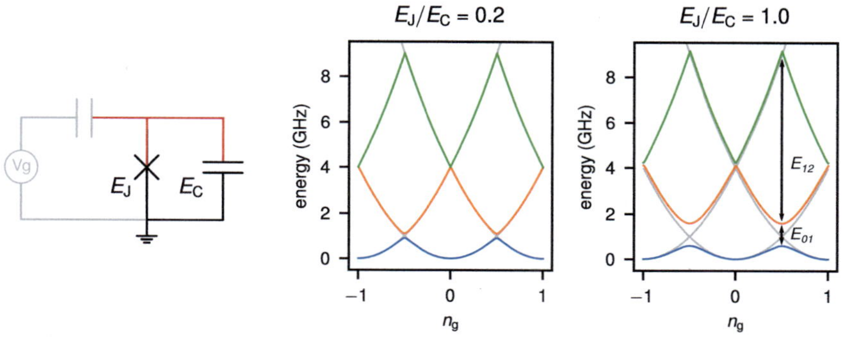

Figure 1.1: Circuit diagram of the Cooper-pair box (CPB) and its eigenenergies. The CPB consist of a superconducting island (indicated in red), which is connected to the outside via a Josephson junction with energy E_J. Cooper-pairs can tunnel via the Junction on and off the island, and their transfer is associated with the charging energy E_C, due to the circuit capacitance. Introducing a gate capacitor connected to a voltage source V_g allows to model the influence of fluctuating charges on the CPB. Increasing the ratio E_J/E_C leads to smaller variations of the transition energies, at the cost of a more harmonic energy level diagram. In the case $E_J/E_C \sim 50$, the charge dispersion is strongly reduced, but also the anharmonicity is only a few percent. Indicated in grey are the quadratic eigenenergies of the Hamiltonian without Josephson coupling.

However, as the ratio E_J/E_C is increased, the change of the transition energies with n_g is exponentially suppressed proportional to $\sqrt{E_J/E_C}$ [8]. In this regime, the circuit is called a *transmon* [8], which currently is the predominantly used superconducting qubit in academia and also in industrial pursuits of a quantum computer. As already mentioned, the transmon's susceptibility to charge noise is strongly suppressed, by adding a shunting capacitor to the Cooper-pair box circuit, which decreases E_C. The price one has to pay for this charge noise insensitivity is a greatly reduced anharmonicity compared to the CPB, which for typical transmons is on the order of 250 MHz at operating frequencies $\sim$ 5 GHz.

Over the last decade the transmon qubit has enabled an increase in coherence times of 3 to 4 orders of magnitude and has been used to show a number of hallmark proof of principle experiments towards the realization of quantum computing in superconducting circuits

[22–24]. One of the major challenges in scaling up to a larger number of qubits is the already mentioned small anharmonicity, which could lead to frequency crowding for larger systems and also potentially result in single qubits leaving the computational subspace of the transmon level structure. A promising alternative for the transmon qubit, albeit currently only at the level of single qubits, could be circuits with high characteristic impedance. The most widely known example, the fluxonium qubit, will be discussed in the following section.

1.2 The fluxonium qubit

1.2.1 Quantum fluctuations of charge and phase

Before diving into the details of the fluxonium qubit Hamiltonian, first we shall discuss the term *high characteristic impedance* and give it a quantitative meaning. To gain insight into circuits with high characteristic impedance, we first look at a the most basic superconducting quantum circuit, the harmonic *LC* oscillator, which is described by the Hamiltonian

$$\hat{H} = \frac{1}{2C}\hat{Q}^2 + \frac{1}{2L}\hat{\Phi}^2. \tag{1.3}$$

Following the standard procedure of expressing the harmonic oscillator Hamiltonian in terms of creation $\hat{a}^\dagger$ and annihilation operators $\hat{a}$ we find the following equations [25]

$$\hat{\Phi} = \Phi_{\mathrm{ZPF}}\left(\hat{a}+\hat{a}^\dagger\right), \quad \hat{Q} = -iQ_{\mathrm{ZPF}}\left(\hat{a}-\hat{a}^\dagger\right).$$

The zero point fluctuations of flux $\Phi_{\mathrm{ZPF}} = \sqrt{(\hbar Z)/2}$ and charge $Q_{\mathrm{ZPF}} = \sqrt{\hbar/(2Z)}$ are defined such that the uncertainties of the quantum mechanical ground state in the phase and charge degree of freedom are given by the respective squared zero point fluctuations. Rewriting both uncertainties in terms of their characteristic scale, and introducing an impedance normalized to the resistance quantum $z = Z/R_{\mathrm{Q}} = Z/[h/(2e)^2]$, we find [25]

$$\Phi_{\mathrm{ZPF}} = \Phi_0\sqrt{\frac{z}{4\pi}},$$
$$Q_{\mathrm{ZPF}} = 2e\sqrt{\frac{1}{4\pi z}},$$

with the flux quantum $\Phi_0 = h/(2e)$. An interesting consequence of these expressions is that for circuits with a characteristic impedance larger than R_Q their quantum fluctuations of the charge degree of freedom are suppressed in favor of larger fluctuations of the flux degree of freedom. For that reason, we call an impedance Z larger than R_Q *high* characteristic impedance.

1.2.2 The uncoupled fluxonium Hamiltonian

As already touched on in Sec. 1.1, the tremendous popularity of the transmon qubit can be attributed to its insensitivity to charge noise, the simplicity of the circuit, and the availability of analytical formulae, which approximate the characteristics of the Hamiltonian very well [8]. Nevertheless, one of the main drawbacks of transmon qubits is their low anharmonicity, which potentially limits the speed of single qubit gates. A promising alternative superconducting qubit is the so called fluxonium. This circuit was invented in 2009 [15] and recently gained increasing popularity [17, 18, 26].

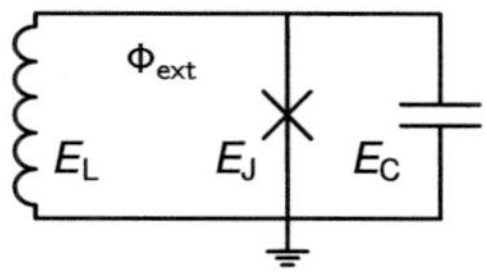

Figure 1.2: Equivalent circuit of the fluxonium qubit. A small Josephson junction with energy E_J is shunted by a large inductor, with characteristic impedance $Z > R_Q$. This so called superinductor adds a characteristic energy E_L to the circuit. The Josephson junction capacitance is lumped together with other capacitances in the circuit and gives rise to a charging energy E_C. See Eq. (1.4) for the circuit Hamiltonian.

Figure 1.2 shows the fluxonium equivalent circuit, which is a small Josephson junction shunted by a large inductor with high characteristic impedance, the so called superinductor. In principle, the fluxonium circuit is comparable in simplicity to the transmon circuit, in practice however the complexity lies in the realization of the superinductor, which needs an inductance on the order of hundreds of nH, while minimizing its parasitic capacitance to ground (cf. Sec. 1.3.1).

Again relying on the circuit quantization technique [21], the bare fluxonium Hamiltonian reads

$$\hat{H} = 4E_C\hat{N}^2 + \frac{1}{2}E_L\hat{\varphi}^2 - E_J\cos\left(\hat{\varphi} - 2\pi\frac{\Phi_{ext}}{\Phi_0}\right), \tag{1.4}$$

where similarly to the CPB Hamiltonian in Sec. 1.1 we have introduced the characteristic energy scales of the circuit, with charging energy $E_C = e^2/(2C)$, inductive energy $E_L = (\Phi_0/2\pi)^2/L$, and Josephson energy $E_J = (\hbar I_c)/2e = (\Phi_0/2\pi)^2/L_J$. Note here an important difference to the previously shown CPB Hamiltonian [see Eq. (1.2)]: the susceptibility of the circuit to any offset charge n_g can be removed through a gauge transformation [27], which can be intuitively understood looking at the circuit diagram. The large inductance provides a shunt to the Josephson junction at low frequencies and the fluxonium circuit possesses no island. Hence, the fluxonium qubit is intrinsically insensitive to charge fluctuations.

Looking at the first two terms of Eq. (1.4), we identify the Hamiltonian of an LC oscillator. Through the correspondence principle we associate the phase degree of freedom with the position and conversely the charge degree of freedom with momentum. In analogy to the harmonic oscillator, the capacitance C of the fluxonium now corresponds to the mass of the effective particle, while the frequency ω of the harmonic oscillator corresponds to $1/\sqrt{LC}$, as expected. Still ignoring the Josephson junction term, we chose the harmonic oscillator eigenstates [28] as the basis for the ensuing numerical diagonalization of the Hamiltonian [29, 30],

$$\psi_n(\varphi) = \frac{1}{\sqrt{b}} \frac{1}{\sqrt{2^n n! \sqrt{\pi}}} H_n\left(\frac{\varphi}{b}\right) e^{-\frac{1}{2}\left(\frac{\varphi}{b}\right)^2}, \tag{1.5}$$

with $b = \left[(8E_C\hbar^2)/E_L\right]^{1/4}$, and H_n the Hermite polynomials. Taking into account the full fluxonium Hamiltonian again, we can compute the matrix elements as follows,

$$\langle \psi_m(\varphi)| \hat{H} |\psi_l(\varphi)\rangle = \hbar\omega \left(m + \frac{1}{2}\right) \delta_{m,l} - E_J \left\langle \psi_m(\varphi) \middle| \cos\left(\hat{\varphi} - 2\pi\frac{\Phi_{ext}}{\Phi_0}\right) \middle| \psi_l(\varphi) \right\rangle. \tag{1.6}$$

In order to obtain the matrix elements of the Josephson junction term, we use a trigonometric identity to split the cosine term,

$$cos\left(\hat{\varphi} - 2\pi\frac{\Phi_{ext}}{\Phi_0}\right) = \sin(\hat{\varphi}) \sin\left(2\pi\frac{\Phi_{ext}}{\Phi_0}\right) + \cos(\hat{\varphi}) \cos\left(2\pi\frac{\Phi_{ext}}{\Phi_0}\right). \tag{1.7}$$

Finally, the matrix elements can be calculated by [29, 30],

$$\langle \psi_m(\varphi)|\sin(\hat{\varphi})|\psi_l(\varphi)\rangle \tag{1.8}$$

$$= \begin{cases} 0, & \text{l + m even} \\ \frac{1}{\sqrt{2^{l+m}l!m!}} 2^{min[l,m]}(min[l,m])!(-1)^{(|m-l|-1)/2} b^{|m-l|} e^{-b^2/4} L^{|m-l|}_{min[l,m]}\left(\frac{b^2}{2}\right), & \text{l + m odd} \end{cases} \tag{1.9}$$

$$\langle \psi_m(\varphi)|\cos(\hat{\varphi})|\psi_l(\varphi)\rangle \tag{1.10}$$

$$= \begin{cases} \frac{1}{\sqrt{2^{l+m}l!m!}} 2^{min[l,m]}(min[l,m])!(-1)^{|m-l|/2} b^{|m-l|} e^{-b^2/4} L^{|m-l|}_{min[l,m]}\left(\frac{b^2}{2}\right), & \text{l + m even} \\ 0, & \text{l + m odd} \end{cases} \tag{1.11}$$

using the associated Laguerre polynomials $L_j^k(x)$ to solve the integrals [31].

By expressing the matrix representation of the fluxonium in the harmonic oscillator basis, we can look at the fluxonium potential and its eigenfunctions at different values of the external magnetic flux, which tunes the Josephson potential with respect to the parabolic potential of the superinductor. Figure 1.3 shows the fluxonium potential (black line) and its five lowest eigenfunctions offset by their corresponding eigenenergy. Table 1.1 shows the fluxonium parameters chosen for the illustration of the potential and fluxonium eigenstates.

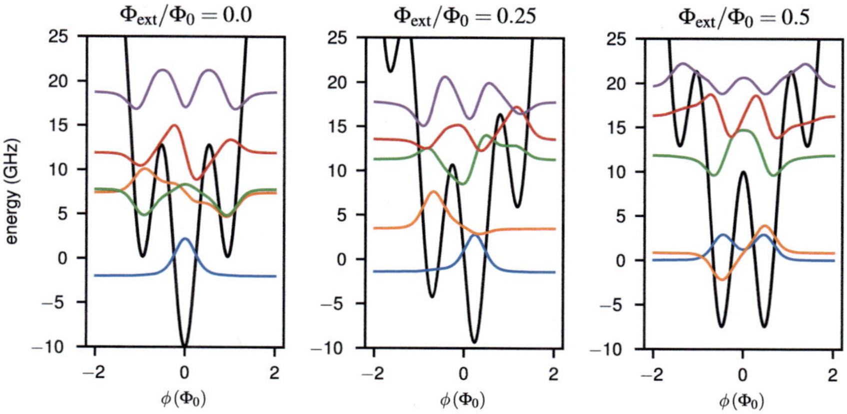

Figure 1.3: Fluxonium potential (black) and eigenstates (colored lines, shifted vertically by their eigenenergy) at different external magnetic flux. By changing the magnetic field threading the fluxonium loop, the potential can be seamlessly tuned from a single well at $\Phi_{ext}/\Phi_0 = 0.0$ to a double well potential at $\Phi_{ext}/\Phi_0 = 0.5$, where the two lowest eigenstates form a well separated effective two-level system. The circuit parameters of the diagonalized Hamiltonian are $E_J/h = 10\,\text{GHz}$, $E_L/h = 0.55\,\text{GHz}$, and $E_C/h = 3.87\,\text{GHz}$.

Table 1.1: Circuit parameters of the fluxonium Hamiltonian plotted in Fig. 1.3.

E_J/h	L_J	E_L/h	L	E_C/h	C
10.00 GHz	16.3 nH	0.55 GHz	300.0 nH	3.87 GHz	5 fF

At zero external flux, we observe a potential with a central well, in which the lowest eigenstate is localized. The three next higher states are localized across three wells, formed by the parabolic superinductor potential modulated by the Josephson junction term. Finally, the highest state shown closely resembles an harmonic oscillator eigenstate, justifying the choice of basis for numerical diagonalization of the fluxonium Hamiltonian.

For an external flux $\Phi_{ext}/\Phi_0 = 0.25$, the central panel in Fig. 1.3 shows how the potential approaches a double well, with the second eigenstate moving down in energy towards the ground state. At $\Phi_{ext}/\Phi_0 = 0.5$ the fluxonium circuit exhibits a symmetric double well potential with two degenerate eigenstates in the left and right well, which hybridize to form an effective two-level system. As is clearly visible these two states are well separated in energy from the higher states, illustrating the large anharmonicity of the fluxonium circuit, while the circuit is still insensitive to charge noise.

Figure 1.4 shows the transition frequencies between the ground and three next higher states as a function of the external magnetic flux for the circuit parameters given in Tab. 1.1. Between the maximum of the lowest transition frequency at $\Phi_{ext}/\Phi_0 = 0.0$ and the minimum at $\Phi_{ext}/\Phi_0 = 0.5$, the frequency can be tuned over ~ 10 GHz (blue line). At $\Phi_{ext}/\Phi_0 = 0.5$, the usual operation point of the fluxonium qubit the spectrum shows an anharmonicity of about a factor of 10 between the lowest transition frequency and the transition to the second excited state. This anharmonicity is about two orders of magnitude larger compared to typical transmon qubits and illustrates the strong isolation of the computational basis from higher excited states in the fluxonium qubit, which strongly decreases leakage out of the qubit subspace.

Different groups have presented energy relaxation times in the range of ms in fluxonium qubits. However, to achieve this two different approaches were employed. Figure 1.5 compares the two potentials and corresponding eigenstates close to the $\Phi_{ext}/\Phi_0 = 0.5$ qubit operating point. The left panel shows the Hamiltonian diagonalization of the device presented in Ref. [16] with characteristic energy scales $E_J/h = 10.2$ GHz, $E_L/h = 0.46$ GHz, and $E_C/h = 3.6$ GHz. In this device T_1 times up to 8 ms were measured thanks to optimization of the fabrication process, the use of a 3D cavity as the readout resonator, which significantly decreases dielectric losses as already presented in transmon qubits [32], and a strongly shielded measurement setup.

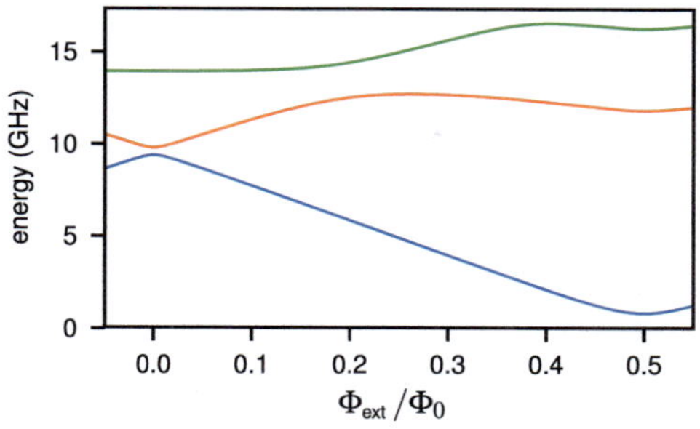

Figure 1.4: Transition frequencies of the fluxonium qubit as a function of the external magnetic flux. Depending on the circuit parameters (cf. Tab. 1.1 for the values used in the plot), the transition frequency can be tuned over several GHz, and the spectrum shows an anharmonicity of a factor ~ 10 at $\Phi_{ext}/\Phi_0 = 0.5$, which is the usual fluxonium operation point.

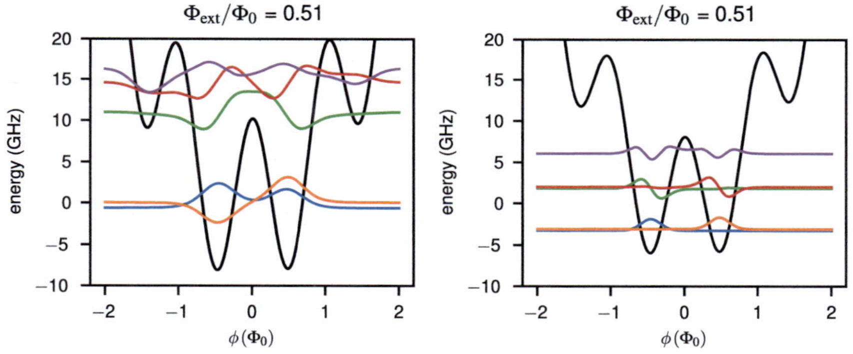

Figure 1.5: Conventional (left, [16]) vs. *heavy* fluxonium (right, [17]) potential and eigenstates. Shunting the fluxonium with a large capacitor $\sim 10\,\mathrm{fF}$ to $10^2\,\mathrm{fF}$ results in strongly localized eigenstates and exponentially suppressed off-diagonal matrix elements [cf. Eq. (1.8) to (1.11)]. Both circuits exhibit energy relaxation times on the order of ms illustrating the potential as well as versatility of the fluxonium circuit, with the heavy fluxonium being operated as a Λ-system.

The right panel of Fig.1.5 shows the potential and eigenfunctions of a so called *heavy fluxonium*, presented in Ref. [17]. Its characteristic energy scales $E_J/h = 8.1\,\mathrm{GHz}$, and $E_L/h = 0.48\,\mathrm{GHz}$ are comparable to those of the fluxonium in Ref. [16], the charging energy $E_C/h = 0.46\,\mathrm{GHz}$ however is significantly decreased by shunting the Josephson junction with a large capacitor. As already mentioned, in analogy to the harmonic oscillator Hamiltonian, in the fluxonium the capacitance C plays the role of the mass, hence the name *heavy* fluxonium. Furthermore, the off-diagonal matrix elements are exponentially suppressed $\propto exp(-E_L/E_C)$, leading to strongly localized eigenstates (cf. right panel Fig. 1.5). Due to this strong localization the operation of the heavy fluxonium relies on cycling the qubit through side transitions. Nevertheless, for the circuit presented in Ref. [17] energy relaxation times on the order of ms were also observed. The comparison between the two fluxoniums serves to illustrate the tremendous potential in terms of qubit coherence, as well as the versatility of the fluxonium circuit in exploring different parameter regimes.

1.2.3 Fluxonium qubit loss channels

To quantify the losses limiting the fluxonium qubit T_1, which will later on serve as a diagnostic tool to investigate the grAl superinductor, we recall the relaxation rate expressed using Fermi's golden rule [16, 29],

$$\Gamma_{1,j} = \frac{1}{\hbar^2} \left|\langle 0|\hat{A}|1\rangle\right|^2 S_{II}^j(\omega_{01}). \tag{1.12}$$

Here, $\langle 0|\hat{A}|1\rangle$ is the matrix element between the qubit ground and excited state with the operator $\hat{A}$, which couples a decay channel j to the qubit. $S_{II}^j(\omega_{01})$ represents the current spectral density of the noise of channel j at the qubit transition frequency ω_{01} and is expressed by [33],

$$S_{II}^j(\omega_{01}) = \hbar\omega_{01} Re[Y_j(\omega_{01})] \left(\coth\left(\frac{\hbar\omega_{01}}{2k_B T}\right) + 1\right), \tag{1.13}$$

where $Re[Y_j(\omega_{01})]$ is the real part of the admittance causing the relaxation of the qubit. In the following we use this general concept of dealing with different sources of dissipation to calculate the relaxation rates induced on the fluxonium by capacitive, radiative, and inductive loss by finding the corresponding real parts of the admittance [29]. Additionally, dissipation induced by quasiparticle tunneling across the small Josephson junction will be evaluated following the approach detailed in Refs. [34, 35].

Capacitive, inductive, and radiative loss

In accordance with Refs. [16, 36] we model the losses in the fluxonium's capacitors by a parallel resistance R_C. Using the definition of the quality factor of a capacitor, in this case $Q = \omega_{01} C R_C$ [37], we find the admittance of the capacitor

$$Y_C(\omega_{01}) = \left(\frac{Q}{\omega_{01}C} + i\frac{1}{\omega_{01}C}\right)^{-1} = \frac{\omega_{01}C(Q-i)}{Q^2+1} \approx \frac{\omega_{01}C}{Q} - \frac{i\omega_{01}C}{Q^2}. \tag{1.14}$$

From which we take the real part for the calculation of the fluxonium relaxation due to dielectric loss in capacitive elements of the circuit.

In the same fashion we calculate the admittance of an inductor with a resistance in series [37],

$$Y_L(\omega_{01}) = \left(\frac{\omega_{01}L}{Q} - i\omega_{01}L\right)^{-1} = \frac{Q(\omega_{01}L + i\omega_{01}LQ)}{\omega_{01}^2L^2 + \omega_{01}^2L^2Q^2} \approx \frac{1}{\omega_{01}LQ} + i\frac{1}{\omega_{01}L}. \tag{1.15}$$

In order to obtain the real part of the admittance as seen from the fluxonium junction, which causes radiative or Purcell loss [38], we employ finite element method simulations. We obtain the frequency dependent admittance by placing a simulation port at the position of the Josephson junction in the 3D model of the fluxonium circuit, which also includes the sample holder. The second simulation port is located on the modeled coaxial line, which connects to the sample holder. Using this approach, the loss through coupling to the outside world is encoded in the admittance seen from the port at the junction position.

With the real parts of the admittance for capacitive, inductive and radiative loss now at hand, the only missing ingredient is the matrix element. Following again the approach presented in Ref. [29], the energy relaxation can be modeled as the phase across the Josephson junction coupling to some source of dissipation. Therefore, the matrix element, which needs to be calculated is (see Fig. 1.6),

$$\langle 0|\hat{\Phi}|1\rangle = \langle 0|(\Phi_0/2\pi)\hat{\varphi}|1\rangle. \tag{1.16}$$

Quasiparticle tunneling induced loss

Despite operating quantum circuits at temperatures $\sim 20\,\mathrm{mK}$, quasiparticle densities normalized to the total number of Cooper-pairs on the order of $x_{QP} \sim 10^{-6}$ have been reported [16, 39–44]. This value, orders of magnitude larger than the thermally expected $x_{QP}^{th} = \sqrt{2\pi k_B T/\Delta_0}\, exp(-\Delta_0/k_B T) \approx 10^{-54}$ in pure thin film aluminum circuits (with superconducting gap $\Delta_0 = 210\,\mu\mathrm{eV}$) [35], has recently been shown to be a non-thermal population [45]. Its source is still under investigation with possible candidates being high-energy particles from local radioactive decay or of cosmic origin [46], or residual photons with energies $hf > 2\Delta_0$, which reach superconducting quantum circuits despite best efforts to shield them from any stray radiation [47].

It has been shown that excess quasiparticles cause energy relaxation and decoherence in superconducting qubits because of their coupling via the Josephson tunneling term [34, 48]. In the following we give the main results of Ref. [34], which allow to connect the normalized excess quasiparticle density in the fluxonium qubit to its energy relaxation time T_1.

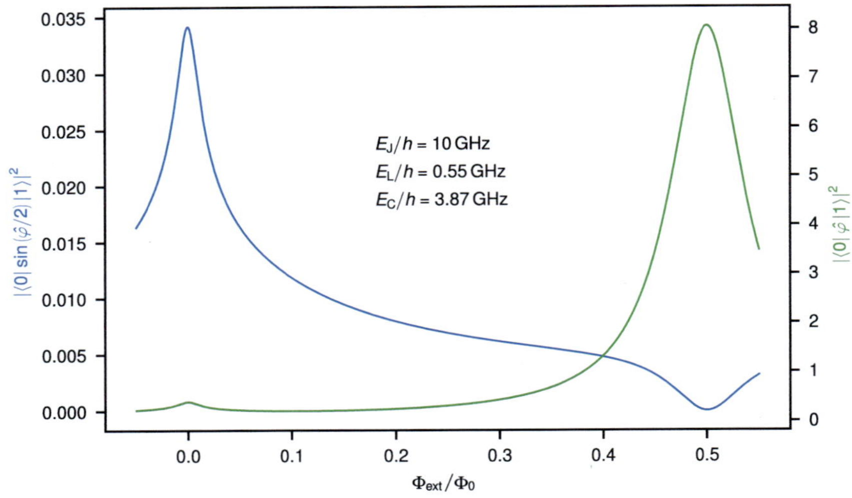

Figure 1.6: Squared absolute of fluxonium matrix elements showing coupling strength to dissipation channels. Blue solid line (left axis) shows $|\langle 0|\sin(\hat{\varphi}/2)|1\rangle|^2$, expressing the coupling between quasiparticles tunneling across the Josephson junction and the fluxonium states. At $\Phi_{\mathrm{ext}}/\Phi_0 = 0.5$ the matrix element vanishes, which is a manifestation of the interference between electron- and holelike quasiparticles, experimentally shown in Ref. [16]. The green solid line (right axis) shows $|\langle 0|\hat{\varphi}|1\rangle|^2$, which is the matrix element used for the calculation of capacitive, inductive, and radiative loss (cf. Tab. 1.2).

Following Catelani *et al.* [34] the transition rate between the first excited and ground state of the qubit can be calculated through Fermi's golden rule,

$$\Gamma_{QP} = \left|\left\langle 0\middle|\sin\left(\frac{\hat{\varphi}}{2}\right)\middle|1\right\rangle\right|^2 S_{QP}(\omega_{01}). \tag{1.17}$$

Here, equal superconducting electrodes of the Josephson junction, and small excitations of the quasiparticles, δE, and qubit, i.e. $\hbar\omega_{01}, \delta E \ll 2\Delta_0$ are assumed. It is already interesting to note that for the fluxonium qubit biased at $\Phi_{\mathrm{ext}}/\Phi_0 = 0.5$ (cf. Fig. 1.3) the matrix element vanishes (see Fig. 1.6). This happens thanks to the symmetric potential, ground state, and $\sin(\hat{\varphi}/2)$ term in Eq. (1.17), and the antisymmetric first excited state at this specific bias point. The effect is a manifestation of the destructive interference between electron- and holelike quasiparticles tunneling across the Josephson junction and has been experimentally verified in Ref. [16]. The resulting insensitivity to dissipation caused by quasiparticle tunneling across the Josephson junction, together with the large anharmonicity, and first order flux noise insensitivity justify the standard operation point of the fluxonium qubit at $\Phi_{\mathrm{ext}}/\Phi_0 = 0.5$.

The normalized quasiparticle current spectral density $S_{QP}(\omega_{01})$, which expresses the quasiparticle kinetics, is found to be [34]

$$S_{QP}(\omega_{01}) = \frac{h\omega_{01}}{\pi e^2} Re[Y_{QP}(\omega_{01})], \quad \text{with} \tag{1.18}$$

$$Re[Y_{QP}(\omega_{01})] = \frac{1}{2} x_{QP} \frac{8E_J e^2}{\Delta_0 h} \left(\frac{2\Delta_0}{\hbar\omega_{01}}\right)^{3/2}. \tag{1.19}$$

Summary of relaxation rates

Table 1.2 summarizes the fluxonium relaxation rates $\Gamma_{1,j} = 1/T_{1,j}$ due to capacitive, inductive, radiative, and quasiparticle loss.

Table 1.2: Relaxation rates from the first excited to the ground state due to capacitive, inductive, radiative and quasiparticle loss in the fluxonium qubit. For the radiative loss contribution the real part of the admittance is obtained through FEM simulations of the circuit geometry in the sample holder, which includes the coupling to the external measurement setup.

Dissipation source	Relaxation rate Γ_1
Capacitive	$\frac{\lvert\langle 0\rvert \hat{\varphi} \lvert 1\rangle\rvert^2}{(2e)^2} \left[\coth\left(\frac{\hbar\omega_{01}}{2k_B T}\right) + 1\right] \frac{\hbar\omega_{01}^2 C}{Q_{\text{cap}}}$
Inductive	$\frac{\lvert\langle 0\rvert \hat{\varphi} \lvert 1\rangle\rvert^2}{(2e)^2} \left[\coth\left(\frac{\hbar\omega_{01}}{2k_B T}\right) + 1\right] \frac{\hbar}{L Q_{\text{ind}}}$
Radiative	$\frac{\lvert\langle 0\rvert \hat{\varphi} \lvert 1\rangle\rvert^2}{(2e)^2} \left[\coth\left(\frac{\hbar\omega_{01}}{2k_B T}\right) + 1\right] \hbar\omega_{01} Re[Y(\omega_{01})]$
Quasiparticle	$\lvert\langle 0\rvert \sin(\hat{\varphi}/2) \lvert 1\rangle\rvert^2 \frac{8E_J}{\hbar\pi} \left(\frac{2\Delta_0}{\hbar\omega_{01}}\right)^{1/2} x_{QP}$

1.2.4 Inductively coupled fluxonium-resonator system

To read out the quantum state of the fluxonium qubit we employ the tools of circuit quantum electro dynamics (cQED), namely the dispersive readout scheme [49, 50]. Concretely, we achieve this by inductively coupling a readout resonator to the fluxonium qubit, as schematically illustrated by the circuit diagram in Fig. 1.7. Besides the bare fluxonium (cf. Fig. 1.2), a readout resonator with inductance L_r and capacitance C_r, as well as the shared inductance L_s between resonator and qubit now make up the circuit.

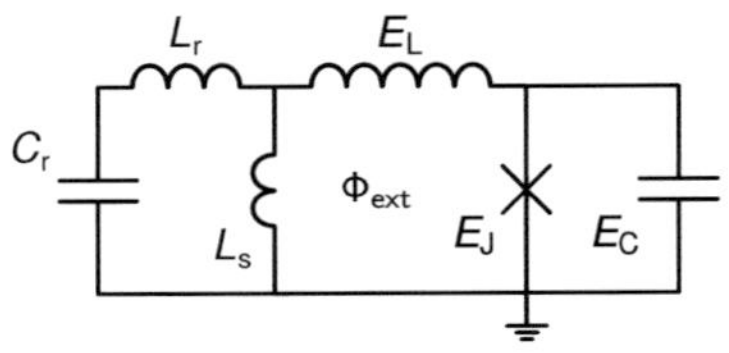

Figure 1.7: Circuit diagram of the inductively coupled fluxonium-resonator system. The fluxonium qubit (E_J, E_L, and E_C, cf. Fig. 1.2) is coupled to a readout resonator (L_r, C_r) via a shared inductance L_s. Depending on the fluxonium qubit state, the resonant frequency of the resonator changes, which is monitored and hence provides information about the qubit state [49, 50].

Using circuit quantization [21], the Lagrangian $\mathscr{L}$ of the fluxonium-resonator system in the limit $L_q \gg L_r, L_s$ is found as detailed in Ref. [30],

$$\begin{aligned}\mathscr{L} = \frac{C_r}{2}\dot{\Phi}_r^2 - \frac{1}{2(L_r+L_s)}\Phi_r^2 + \frac{L_s}{L_q(L_r+L_s)}\Phi_r\Phi_q \\ + \frac{C_q}{2}\dot{\Phi}_q^2 - \frac{1}{2L_q}\Phi_q^2 + E_J\cos\left(2\pi\frac{\Phi_q}{\Phi_0} - \Phi_{ext}\right).\end{aligned} \tag{1.20}$$

Following the steps of decoupling the linearized Lagrangian, and applying a Legendre transformation, as detailed again in Ref. [30], the Hamiltonian of the inductively coupled fluxonium resonator system is found,

$$\hat{H} = \frac{1}{2C_R}\hat{Q}_R^2 + \frac{1}{2L_R}\hat{\Phi}_R^2 + \frac{1}{2C_Q}\hat{Q}_Q^2 + \frac{1}{2L_Q}\hat{\Phi}_Q^2 - E_J\cos\left(2\pi\frac{\lambda_3\hat{\Phi}_R}{\Phi_0} + 2\pi\frac{\lambda_4\hat{\Phi}_Q}{\Phi_0} - \Phi_{ext}\right), \tag{1.21}$$

where λ_3 and λ_4 are chosen such that the linear part of the Lagrangian becomes diagonal. Here, similarly to the solution of the uncoupled fluxonium Hamiltonian, product states of the harmonic oscillator eigenfunctions form the basis for the ensuing numerical diagonalization of the Hamiltonian. The diagonal elements of the matrix are now given by the harmonic oscillator eigenenergies of the resonator [cf. first two terms in Eq. (1.21)], and qubit [terms 3 and 4 of Eq. (1.21)], while the off-diagonal elements are again found using the relations (1.8) - (1.11).

Following numerical diagonalization of the coupled system Hamiltonian, the photon numbers in resonator and fluxonium now have to be assigned to the obtained level structure. This is done by an approach, which to the best of our knowledge has first been shown in Ref. [51]. It relies on the calculation of the transition dipole moments $\langle d_{ij}\rangle$ between the numerically obtained eigenstates of the coupled fluxonium resonator Hamiltonian. Basis for this assumption is the fact that the readout resonator should have a dipole moment significantly larger than the fluxonium, which is calculated by,

$$\langle d_{ij} \rangle \propto \langle \psi_i | \hat{Q}_R + \hat{Q}_Q | \psi_j \rangle \,. \tag{1.22}$$

Starting from the ground state with zero photons in the resonator and fluxonium, one begins climbing the ladder stepping in every iteration to the state with the largest transition dipole moment, always assigning an additional resonator photon. Once this is completed for the states having zero photons in the qubit, the lowest, yet unidentified eigenstate is assigned with having one photon in the fluxonium. Starting from this state, the ladder of adding photons in the resonator is climbed again, to assign the next set of photon numbers. This process continues until all states are assigned a photon number for resonator and fluxonium. Figure 1.8 shows the output of the numerical diagonalization (left) with states simply sorted by their energy, and the result after the just described sorting procedure (right). Assignment of photon numbers to the different lines is now possible, illustrating the effectiveness of the approach invented in Ref. [51].

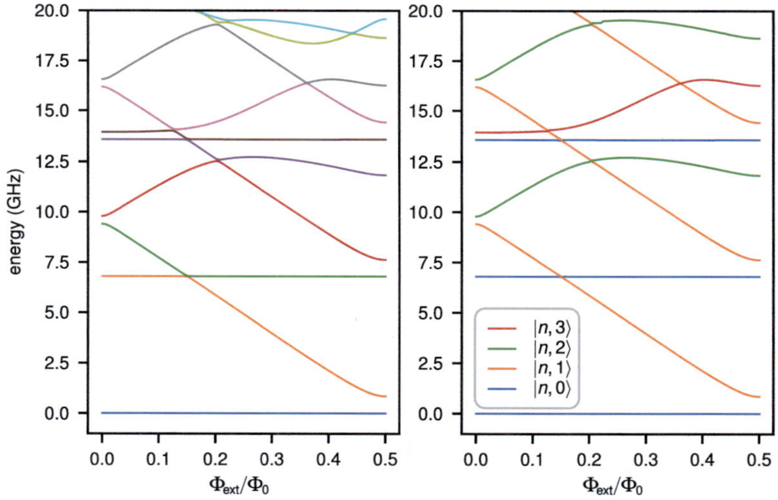

Figure 1.8: Spectrum of the inductively coupled fluxonium resonator system. On the left, levels are sorted by energy. In the right panel states are sorted according to their transition dipole moment (see main text), facilitating the identification of photon number states of fluxonium and resonator. The legend indicates the number of photons in the fluxonium for the different levels of same color, whereas the photon number in the resonator increases by one between every two lines of same color.

With the spectrum of the coupled system now sorted according to photon numbers, one can extract the dispersive shift of the resonator χ caused by the fluxonium state,

$$\chi = \left(E_{|n+1,1\rangle} - E_{|n,1\rangle}\right) - \left(E_{|n+1,0\rangle} - E_{|n,0\rangle}\right). \tag{1.23}$$

Figure 1.9 shows the dispersive shift as a function of the external magnetic flux. As can be seen, it exhibits a rich behavior with changing sign and significantly increased values at positions where crossings between resonator-like and qubit-like eigenstates of the Hamiltonian occur.

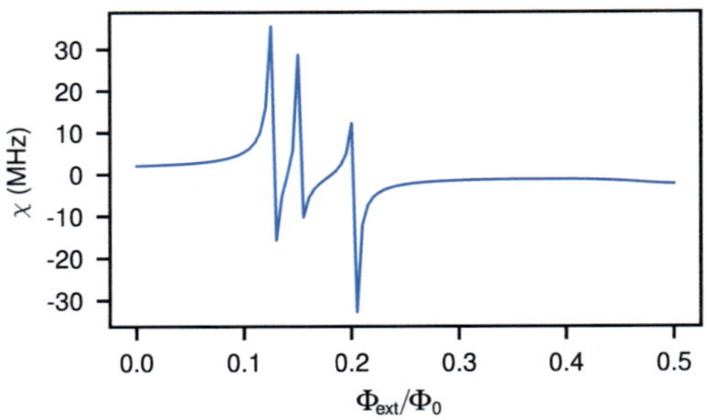

Figure 1.9: Dispersive shift of the readout resonator [cf. Eq. (1.23)]. Based on the sorted eigenstates of the inductively coupled resonator fluxonium Hamiltonian (see Fig. 1.8) we extract the dispersive shift χ. As illustrated by the plot, χ shows a rich behavior, with changing sign and significantly varying magnitude. The dependence of χ on the external magnetic flux strongly depends on the choice of circuit parameters and large values of the dispersive shift indicate crossings between resonator-like and qubit-like states.

1.3 Granular aluminum for high impedance environments

1.3.1 Status quo of high impedance environment realization

From a design point of view three key properties of superinductors are of interest: (i) kinetic inductance, enabling high characteristic impedances, and allowing to add large shunting inductances into superconducting circuits, (ii) behavior as linear inductors, minimizing the addition of nonlinearity into a quantum circuit, (iii) sufficiently low microwave frequency losses avoiding to limit the qubit coherence. In the following we will discuss implementations of superinductors harnessing the kinetic inductance of two types of structures: on the one hand Josephson junction arrays, on the other hand disordered superconductors, mainly different types of nitrides.

Josephson junction chain superinductors

Since Josephson junctions have already been a basic building block of superconducting qubits, which have repeatedly demonstrated remarkable times up to hundreds of micro-seconds [2], the idea to use arrays of large Josephson junctions to realize superinductors naturally arises. A large body of literature exists, which investigates Josephson junction arrays (JJAs) from different viewpoints. Among others, JJAs are specifically investigated serving as superinductors [29, 36, 52, 53], as high impedance environments [54, 55], as resonators with engineered dispersion relation [56, 57], and quantum limited amplifiers with large saturation power [58]. In the following, we will discuss some considerations for the design of superinductors using Josephson junctions and give a brief summary of reported performances.

Due to their relatively easy fabrication by shadow evaporation processes [59–61] and remarkable coherence properties, practically all Josephson junctions in superconducting quantum circuits are realized as $Al/AlO_x/Al$ tunnel junctions, in which the insulating barrier is obtained by a controlled oxidation of the first pure aluminum electrode. According to Kleinsasser *et al.* [62] the critical current density of tunnel junctions with AlO_x barriers is proportional to the product of oxidation pressure p_{ox} and time t_{ox}, the so called exposure $\mathscr{E}_{ox} = p_{ox} \times t_{ox}$. Using the Josephson equations [19] one can furthermore calculate the Josephson inductance $L_J = (\hbar I_c)/2e$, associated with the inertia of Cooper pairs tunneling through such a contact. Owing to the fact that $Al/AlO_x/Al$ Josephson junctions are two superconducting electrodes separated by a thin dielectric layer, they form a parallel plate capacitor with a characteristic capacitance $\mathscr{C}_J$ on the order of $50\,fF/\mu m^2$ to $100\,fF/\mu m^2$ [63–66], which is in parallel with the Josephson inductance L_J (cf. the resistively and capacitively shunted junction (RCSJ) model, presented for example in Ref. [67]). This parallel arrangement of a capacitance and inductance forms an LC-resonator with a characteristic resonant frequency $\omega_p = 1/\sqrt{C_J L_J}$, called the plasma frequency. Let us now examine the dependence of the plasma frequency on design parameters of the Josephson junction, such as junction area A and critical current density.

$$\omega_p = 1/\sqrt{C_J L_J} \tag{1.24}$$

$$= \sqrt{\frac{1}{A\mathscr{C}_J} \times \frac{2\pi I_c}{\phi_0}} \tag{1.25}$$

$$\propto \sqrt{\frac{2\pi \mathscr{E}_{ox}}{\mathscr{C}_J \phi_0}} \quad \text{with } I_c/A \propto \mathscr{E}_{ox} \text{ (cf. Ref. [62]).} \tag{1.26}$$

As briefly discussed earlier, superconducting quantum circuits are commonly operated in the low GHz regime up to values $\sim$10 GHz. This means a superinductor as a fundamental building block of such a circuit should not possess a self-resonance in this frequency range, meaning $\omega_\mathrm{p}/2\pi \gtrsim 10\,\mathrm{GHz}$. Since the plasma frequency is independent of the junction area and only dependent on the critical current density of the junction, i.e. the oxidation exposure [cf. Eq. 1.26], this places a stringent design bound on the Josephson junctions usable for a superinductor.

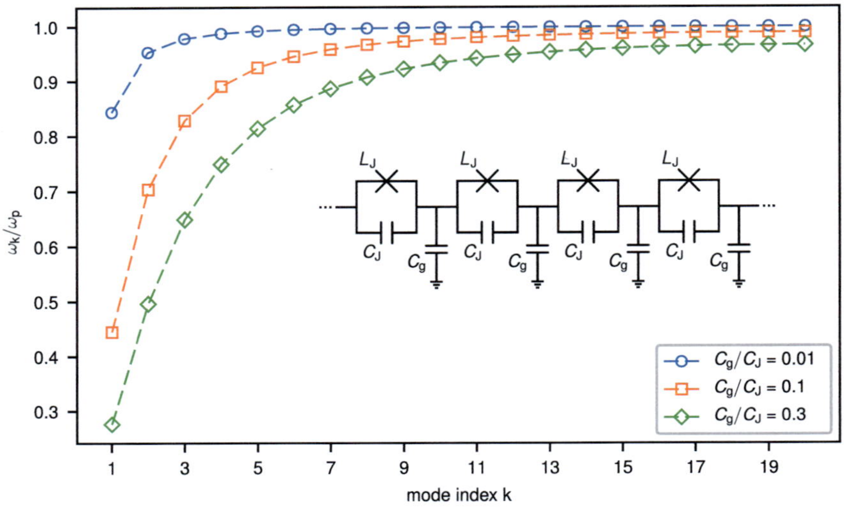

Figure 1.10: Circuit model of an array of Josephson junctions and its dispersion relation for different ratios between Josephson capacitance C_J and capacitance to ground C_g. To form the superinductor N junctions are connected in series to increase the total inductance and maintain a high characteristic impedance. For illustration purposes, in the plot we have chosen $N = 20$. The Josephson inductance L_J and capacitance C_J of the individual junctions set the plasma frequency $\omega_\mathrm{p} = 1/\sqrt{L_\mathrm{J}C_\mathrm{J}}$, which only depends on the critical current density of the junction and is independent of the junction area (see main text for a more detailed discussion). This plasma frequency sets the upper bound for all resonant modes of the junction array, which has a dispersion relation $\omega(k) = \omega_\mathrm{p}\sqrt{[1-\cos(k\pi/N)]/[1-\cos(k\pi/N)+C_\mathrm{g}/(2C_\mathrm{J})]}$, where k is the mode index [36, 68]. Depending on the ratio $C_\mathrm{g}/C_\mathrm{J}$, the dispersion relation changes, with lower resonant frequencies of the first array modes as $C_\mathrm{g}/C_\mathrm{J}$ increases (C_J kept constant for all curves).

Furthermore, to fulfill the requirement of a linear superinductor, the phase across the Josephson junctions making up the array should be small, to permit the expansion of the cosine to lowest order only. Consequently, the junctions of the superinductor have larger critical currents, which correspond to Josephson inductances $\sim 3\,\mathrm{nH}$ [16]. In order to realize a superinductor only an array of $N \sim 10^2$ Josephson junctions can fulfill all requirements simultaneously. Since any length of metallic wire is also associated with a capacitance to its surroundings, one now not only has to take into account the plasma frequency of the individual Josephson junctions forming the array, but also the collective modes of the complete array. This means that additionally to the Josephson capacitance any capacitance of the individual metal islands to ground has to be taken into account. For the same reason as for the plasma frequency, all array modes should be above the operating frequency of the quantum circuit employing the superinductor. A circuit quantization treatment [21] of the model for a Josephson junction chain with capacitance to ground C_g (see inset in Fig. 1.10), presented in Refs. [36, 68], leads to the following dispersion relation

$$\omega(k) = \omega_p \sqrt{\frac{1-\cos(k\pi/N)}{1-\cos(k\pi/N)+C_g/(2C_J)}}. \tag{1.27}$$

From Eq. 1.27 we see, that all N modes of the Josephson junction array are distributed below the plasma frequency. This can be intuitively understood by imagining that for the highest array resonance neighboring islands charge with opposite sign, with the charge concentrated in the parallel plate capacitor of the Josephson junction, only renormalized by the capacitance to ground C_g of the islands. For the low frequency spectrum however, Fig. 1.10 shows the significant influence of C_g assuming a constant C_J. Imagining a junction chain with two open boundary conditions, the fundamental mode has a $\lambda/2$ field distribution, where the capacitive loading of the line by the capacitance to ground C_g lowers the resonant frequency. Assuming a plasma frequency $\omega_p/2\pi = 20\,\mathrm{GHz}$ and a ratio $C_g/C_J = 0.1$ (cf. orange squares in Fig. 1.10), this leads to a lowest resonant frequency of the array on the order of 9 GHz. Since all self resonances of the superinductor should be well above the qubit operating frequency, this example illustrates the difficulty in designing a superinductor from an array of Josephson junctions, which simultaneously fulfills the constraints: $Z > R_Q$, $L \sim 10^2\,\mathrm{nH}$, linear inductor behavior, and lowest resonant frequency $\gtrsim 10\,\mathrm{GHz}$.

Table 1.3: Overview of properties of Josephson junction array superinductors evaluated independently (Refs. [36, 52]) or implemented in fluxonium qubits (Refs. [16, 18]).

Z/R_Q	L (nH)	Q_i	T_1 (μs)	Ref.
0.7	150	5×10^4	-	N. A. Masluk et al. [36]
3.8	3000	-	> 1	M. T. Bell et al. [52][c]
3.1	330	$\sim 3 \times 10^6$	$\gtrsim 10^3$	I. M. Pop et al. [16]
0.6 - 1.4	$72 - 430$	-	$70 - 270$	L. B. Nguyen et al. [18]

[c] It should be noted this superinductor consists of multiple unit cells of asymmetric SQUIDS, where two adjacent cells share a large Josephson junction. The values quoted in the table are achieved at full SQUID frustration, i.e. an external magnetic flux of $\Phi_0/2$ threading the loops of the arrays. At integer values of the external flux, the inductance decreases down to 16 nH. Therefore, to operate this type of superinductor in a fluxonium it needs a local flux control for the superinductor to enable independent tuning of the fluxonium qubit transition frequency.

Table 1.3 shows exemplary benchmark results of Josephson junction array superinductors reported in the literature to provide a comparison for the performance of granular aluminum superinductors. We use the results of four publications involving junction array superinductors, and give their impedance normalized to the resistance quantum R_Q, their total inductance, internal quality factors, and energy relaxation times of fluxonium qubits realized with such superinductors. As can be seen, Josephson junction arrays have been successfully used to implement fluxonium qubits with remarkable energy relaxation times $T_1 \gtrsim 1\,\mathrm{ms}$. It should be noted however, that Ramsey fringes and spin echo coherence times of fluxonium qubits were only reported in the 10's of microseconds range, two orders of magnitude smaller than their T_1 times. Only very recently $T_2 \gtrsim 100\,\mu\mathrm{s}$ [18] were reported in fluxonium qubits. Further investigation into the limiting decoherence mechanisms in fluxonium qubits has to be performed to elucidate and mitigate noise sources severely reducing T_2 below the theoretically attainable limit of $T_2 \leq 2T_1$.

Finally, we examine the nonlinearity of Josephson junction chains in terms of the quantum optics Hamiltonian [69],

$$\hat{H}/\hbar = \sum_{n=1} \left(\omega_n + K_{nn}\hat{a}_n^\dagger \hat{a}_n\right) \hat{a}_n^\dagger \hat{a}_n + \sum_{n,m=1;n\neq m} \frac{K_{nm}}{2} \hat{a}_n^\dagger \hat{a}_n \hat{a}_m^\dagger \hat{a}_m. \qquad (1.28)$$

In this framework ω_n are the mode frequencies with corresponding bosonic creation and annihilation operators $\hat{a}_n^\dagger$ and $\hat{a}_n$. The first order nonlinearity of the n-th mode is accounted for by the self-Kerr coefficient K_{nn}, which describes the frequency shift of a mode as a function of its occupation number. Similarly, the interaction between modes is expressed in terms of the cross-Kerr coefficient K_{nm}, describing how much the frequency of mode n changes as a function of the occupation of mode m. In Josephson junction array superinductors, self- and cross-Kerr coefficients are on the order of a few MHz/photon [36, 56].

Disordered superconductors for superinductors

Due to their high kinetic inductance, disordered superconductors are a natural approach for the realization of superinductors. Different types of nitrides are the most often used materials. However, they have only recently received attention in the field of superconducting quantum circuits. Table 1.4 gives an incomplete overview of different superconducting circuits realized with disordered superconductors. Due to their relatively high kinetic inductance, which leads to a larger signal for a given incident power in kinetic inductance detectors (KID), a lot of the research has been performed in the context of microwave kinetic inductance detectors.

As the table shows, most implementations posses moderate kinetic inductances of up to 0.2 nH/□, which mostly do not put the circuit's characteristic impedances above the resistance quantum R_Q. Additionally, most of the obtained results are measured at readout powers, where the average number of photons circulating in the resonator $\bar{n} \gg 1$. Due to the saturation of different loss mechanisms at higher readout powers, the measured internal quality factors of these materials are higher, making it difficult to compare between the different publications and precisely evaluate their potential for superconducting qubits. Notable exceptions are Refs. [26, 70], where disordered superconductors were specifically investigated for the purpose of superinductors in qubits. The kinetic inductances obtained in these experiments are in the range of 0.08 nH/□ to 0.2 nH/□, while internal quality factors in the single photon regime are $\lesssim 5 \times 10^4$. Since both approaches minimized the superinductor's stray capacitance by employing a nanowire geometry with width $w \lesssim 100$ nm the characteristic impedances of the superinductors exceed the resistance quantum. From a fabrication point of view, disordered superconductors, especially different types of nitrides necessitate complex film deposition processes with stringent requirements for substrate temperature during deposition, exact film growth rate controls, etc. Despite these challenges, very high internal quality factors have been achieved in quantum circuits implemented with disordered superconductors [71, 72].

Table 1.4: Overview of properties of superconducting circuits employing disordered superconductors.

Material	$L_\square$ (nH/□)	Q_i	Ref.
TiN	~ 0.1	$\sim 10^6$	[71, 73, 74]
NbTiN	~ 0.1	$\sim 10^5$	[26, 72, 75]
NbN	~ 0.1	$\sim 10^5$	[70, 76, 77]

As already discussed, the added nonlinearity of a superinductor should be as small as possible (cf. discussion in Sec. 1.3.1). In superconducting wires the nonlinearity arises from the dependence of the kinetic inductance on the current flowing in the wire [78]

$$L_\mathrm{k}(I) = L_\mathrm{k}(0)\left[1 + \left(\frac{I}{I_*}\right)^2\right], \tag{1.29}$$

where $L_\mathrm{k}(0)$ can be approximated using Eq. (1.35), and I_* is comparable to the critical current of the wire $\sim I_\mathrm{c}$ [78]. Since superconducting qubits are operated in the low power regime, where the circulating current is much smaller than the critical current, the first order nonlinearity introduced by the inductance of disordered superconductors is in the range of kHz/photon expressed in terms of the self-Kerr coefficient [cf. Eq. (1.28)] [79]. This is approximately three orders of magnitude smaller than in Josephson junction arrays.

1.3.2 Kinetic inductance and nonlinearity of granular aluminum

As described in Section 1.2.1 high impedance quantum circuits require characteristic impedances $Z > R_Q$ to suppress zero point fluctuations of the charge. A rough estimate of the impedance achievable by simply using the geometric inductance of a wire while minimizing the capacitance, for example in a microstrip geometry, leaves one with an impedance on the order of $Z = 0.2\,\mathrm{k\Omega}$ assuming a 330 µm thick c-plane sapphire substrate and a strip width of 1 µm [d]. A ten times narrower strip of 100 nm width already necessitates a fabrication process involving electron-beam lithography, and results in an impedance $Z \approx 0.25\,\mathrm{k\Omega}$ or $\sim 4\%$ of R_Q. These two estimates already indicate the need for added inductance, aside from the geometrical one, to achieve significantly higher impedances in the range of R_Q. Recently however, fabrication advances have enabled the implementation of a superinductor based on the geometric inductance of a long, thin wire, wound in a spiral and placed on a very thin suspended membrane [80]. Impressive as it is, the complex, multi-layer fabrication process for this geometric superinductor approach significantly complicates the realization of high impedance quantum circuits.

Conveniently, in superconductivity the inertia of Cooper pairs gives rise to the so called *kinetic inductance*, which can be several times larger than the geometric inductance, without adding any spurious capacitance. Recalling the Drude model and assuming a general two-fluid model, where the total number of electrons in a conductor can be divided into a superconducting part n_s and a normal conducting part n_n, the response of a superconductor to an alternating electric field $\vec{E}\exp(i\omega t)$ can be calculated [67].

$$\vec{J} = \left(\sigma_{1,j}(\omega) - i\sigma_{2,j}(\omega)\right)\vec{E} \tag{1.30}$$

$$\sigma_j = \frac{n_j e^2 \tau_j}{m(1+\omega^2\tau_j^2)} - i\frac{n_j e^2 \omega \tau_j^2}{m(1+\omega^2\tau_j^2)} \tag{1.31}$$

Here, the index j discriminates between normal conducting (n) and superconducting electrons (s), and n is the density of electrons, e stands for the electron charge, τ is the electron scattering time, m the electron mass, and ω the frequency of the AC electric field. For superconductors at temperatures $T \ll T_\mathrm{c}$, for which $\tau_s \to \infty$, and frequencies ω well below the superconducting energy gap, Eq. (1.30) reduces to [81]

[d] Impedances are estimated using a numerical approximation for microstrip lines described for example in Ref. [37].

$$\vec{J} \approx -i\sigma_{2,s}\vec{E}. \tag{1.32}$$

From this, it is clear that for any non-zero frequency of an alternating electric field the superconducting electrons don't provide a perfect shunt to the normal conducting electrons. Hence, even small numbers of unpaired electrons, so-called excess quasiparticles add dissipation to quantum circuits and potentially limit the coherence of superconducting qubits, as will be discussed in later chapters. Still, the surface impedance of a thin superconductor with thickness t is almost purely inductive due to the vanishing resistance [81]

$$Z_\mathrm{s} = 1/(\sigma_1 - i\sigma_2) = R + i\omega L_\mathrm{k} \tag{1.33}$$

$$L_\mathrm{k} = \frac{1}{\omega\sigma_2 t} \tag{1.34}$$

Using the Mattis-Bardeen formula in the low temperature limit ($T \ll T_\mathrm{c}$) [67], and employing the definition of the normal state conductivity $\sigma_n = \rho^{-1} = [Rt(w/l)]^{-1} = (R_\square t)^{-1}$ for a conductor of thickness t, width w, and length l, yields a handy approximation for the kinetic inductance per square as a function of its normal state sheet resistance per square:

$$L_\mathrm{k} = \frac{\hbar R_\square}{\pi\Delta_0} \tag{1.35}$$

In order to achieve the high kinetic inductance needed to realize superinductors with impedances exceeding the resistance quantum R_Q, materials with a large normal state resistance per square $R_\square$ are desirable. Granular aluminum is a material well suited for this purpose since one of the key property of grAl is the large tunability of its resistivity ρ which allows the deposition of superconducting films with $\rho = 1\,\mu\Omega\,\mathrm{cm}$ to $\rho = 10^4\,\mu\Omega\,\mathrm{cm}$ before the material undergoes a superconductor to insulator transition [82, 83] (see Fig. 1.11). As grAl films are simply fabricated by depositing pure aluminum in a slight oxygen atmosphere (cf. Chapter 2 and Appendix A.1), the film's resistivity, which controls the kinetic inductance of the wires, can be adjusted by controlling the oxygen partial pressure during grAl film deposition. The large tuning range of ρ as shown in Fig. 1.11(a), adjustable over 4 orders of magnitude, combined with control of the film thickness, allow a fine grained adjustment of the sheet resistance per square $R_\square$, and consequently of the kinetic inductance of a grAl wire.

Figure 1.11(b) shows a chip with three rectangular strip grAl resonators. The dimensions of the resonators, patterned on a sapphire substrate without any backside metalization (cf. Sec. 1.4 for more details on the sample geometry), are $400\,\mu\mathrm{m} \times 5.4\,\mu\mathrm{m}$ (left), $1000\,\mu\mathrm{m} \times 40\,\mu\mathrm{m}$ (middle), and $600\,\mu\mathrm{m} \times 10\,\mu\mathrm{m}$, with corresponding frequencies of 6.994 GHz, 6.025 GHz, and 6.287 GHz. By comparing these results to finite elements methods (FEM) simulations, we extract a kinetic inductance $L_k = 2\,\mathrm{nH}/\square$ for the $\rho = 4 \times 10^3\,\mu\Omega\,\mathrm{cm}^2$, 20 nm thick film. Using Eq. 1.35, we expect a kinetic inductance of $L_k = 1.5\,\mathrm{nH}/\square$ assuming a superconducting gap of 290 μeV [46], which is 25% smaller than concluded from FEM simulations. The reason for this could be on the one hand a geometry of the fabricated samples, which differs from the designed one and hence causes an overestimation of the number of squares, which results in a larger kinetic inductance per square. On the other hand, recent results show that the ratio between the superconducting gap and the critical temperature $\Delta/k_B T_c$ differs from the conventional BCS result of 1.764 and is increased to ~ 2.2 [82, 84], which could also result in a correction factor for the Mattis-Bardeen formula explaining the observed discrepancy between the calculated and simulated kinetic inductance per square.

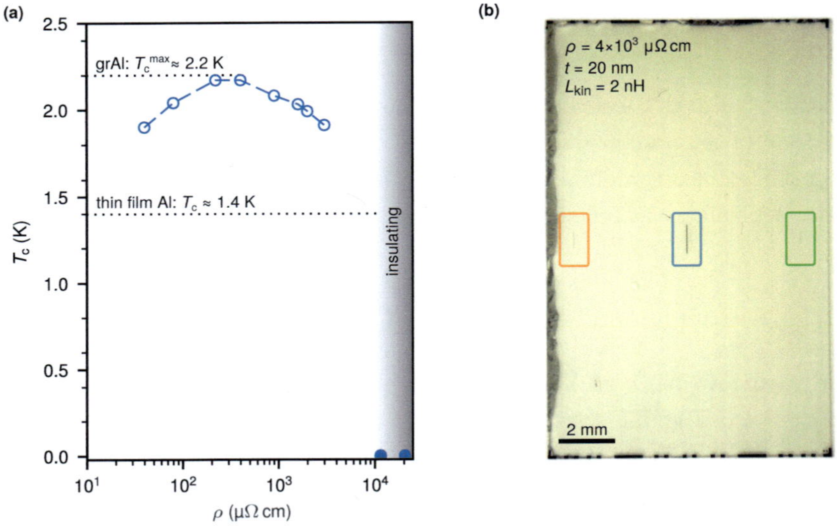

Figure 1.11: Critical temperature of grAl as a function of the room temperature resistivity. The critical temperature of grAl thin films deposited on a room temperature substrate shows a dome shape with T_c (open circles) increasing up to $T_c^{max} \approx 2.2\,\mathrm{K}$ at $\rho \sim 4 \times 10^2\,\mu\Omega\,\mathrm{cm}$. At a resistivity on the order of $10^4\,\mu\Omega\,\mathrm{cm}$ grAl undergoes a superconductor to insulator transition (cf. filled circles). Dashed lines are a guide to the eye, connecting the points. The plot is reproduced according to the results presented in Ref. [82], and based on measurements of samples fabricated at KIT (cf. Appendix A.1 for a detailed description of the fabrication process). The results are in agreement with previous reports about the grAl phase diagram obtained by Deutscher et al. in 1973 [83].

As discussed in the previous paragraph, at high resistivities, granular aluminum reaches kinetic inductances per square approximately a factor of $10-20$ larger than demonstrated in other disordered superconductors (cf. Tab. 1.4). If converted into a kinetic inductance per length, granular aluminum wires exhibit a comparable or even larger kinetic inductance per length than Josephson junction array superinductors (cf. Tab 1.3). Therefore, granular aluminum is a promising material for the realization of high-kinetic inductance wires with characteristic impedance larger than the resistance quantum, while maintaining a straightforward fabrication process and in-situ integrability with current conventional Al/AlO_x/Al Josephson junction fabrication techniques [59–61].

In the following, we will briefly discuss the nonlinearity of granular aluminum wires and compare it to the already presented alternatives for the implementation of superinductors. Motivated by the microstructure of grAl, i.e. pure aluminum grains with 3 nm diameter embedded in a non-stoichiometric AlO_x matrix, Maleeva *et al.* devised a model of the low frequency resonances of grAl wires [79]. Due to its microstructure, a reasonable idea is the description of grAl as a 1D array of effective Josephson junctions, much like depicted in Fig. 1.10 (cf. Chapter 2). Since the grain size a of the pure aluminum grains is approximately independent of the resistivity for $\rho > 10\,\mu\Omega\,\mathrm{cm}$, and constant at $(3 \pm 1)\,\mathrm{nm}$, the entire resonator length of a grAl rectangular resonator is divided into 3 nm wide slices. Each of these slices has an effective, identical critical current I_c, capacitance C_J, and capacitance C_0 to ground. A transformation from the discrete circuit model illustrated in Fig. 1.10 to a continuous description allows to calculate the dispersion relation of a rectangular grAl resonator with length l [79], and mode index n,

$$\omega_n = \frac{na\pi}{l}\sqrt{\frac{2eI_c}{\hbar\left(C_0 + \frac{n^2a^2\pi^2}{l^2}C_J\right)}}. \tag{1.36}$$

A further analysis detailed in Ref. [79] yields the self- and cross-Kerr coefficients for the fundamental mode,

$$K_{1n} \approx \pi e a \frac{\omega_1 \omega_n}{j_c V_{\mathrm{grAl}}}, \quad \text{with } n \geq 1, \tag{1.37}$$

where ω_n are the eigenfrequencies, j_c is the critical current density of the film, and V_{grAl} is the total volume of granular aluminum. In grAl, j_c is proportional to $1/\rho$ [84], meaning that as the kinetic inductance of grAl is increased by $L_k \propto \rho$, also its nonlinearity increases. This can be counteracted by increasing the volume of grAl. As shown by a large number of measurements in Ref. [79] (some of which are discussed in Ch. 3), Eq. 1.37 predicts the nonlinearity of granular aluminum within an order of magnitude and the measured self-Kerr in grAl ranges from 2×10^{-2} Hz/photon to 3×10^{4} Hz/photon. Therefore, grAl is a material ideally suited to realize superinductors with characteristic impedance $Z > R_Q$ and total inductance in the range of 10^2 nH, while introducing a small nonlinearity into the circuit only, equivalent to self-Kerr coefficients on the order of Hz/photon [46].

1.4 Loss mechanisms in granular aluminum

Having established that granular aluminum is a suitable material to fabricate circuit elements with high kinetic inductance, and an impedance larger than the resistance quantum, the following section briefly examines the loss mechanisms in grAl.

1.4.1 Resonator measurements to evaluate losses

A straightforward and universal approach to evaluate the losses of a material at microwave frequencies is to incorporate it into a resonator. Commonly, the losses in any sort of resonant system are described by their quality factors, i.e. the inverse of the loss tangent $Q = 1/\tan\delta$, which is defined as $Q = (\omega W_{\mathrm{tot}})/P_{\mathrm{loss}}$ with ω the resonant frequency, W_{tot} the total average energy stored, and P_{loss} the dissipated power [37]. Depending on the resonator geometry, which determines the distribution of electric and magnetic fields of the standing modes, and the area or volume that is filled with the material in question, the resonator exhibits a variable susceptibility to losses caused by the investigated material.

In order to distinguish between losses that arise due to the intentional coupling of the resonator to an external environment, for example the measurement setup, and losses or dissipation mechanisms that are intrinsic to the resonator, one uses an external or coupling quality factor Q_{c} to quantify the former and an internal quality factor Q_{i} for the latter. The total quality factor combines both and is defined as the inverse sum of the coupling and internal quality factors $Q_{\mathrm{l}} = (1/Q_{\mathrm{c}} + 1/Q_{\mathrm{i}})^{-1}$.

Starting from the Norton equivalent of a circuit that models the reflection measurement of a parallel *RLC* resonator, which is connected to a vector network analyzer (VNA) through a coupling impedance $R_{\mathrm{e}} + iX_{\mathrm{e}}$ and a transmission line (cf. Fig. 1.12), we define the coupling coefficient κ as the ratio between the power dissipated outside the resonator, modeled by a resistor R_{ex}, and inside the resonator, described by the parallel resistor R_0. In terms of the initial circuit parameters shown in Fig. 1.12 this reads,

$$\kappa = \frac{R_0}{R_{\mathrm{ex}}} = \frac{R_0(Z_0 + R_{\mathrm{e}})}{(Z_0 + R_{\mathrm{e}})^2 + X_{\mathrm{e}}^2}. \tag{1.38}$$

In practice, we assume a lossless coupling mechanism, i.e. $Re = 0$, as well as a coupling reactance, which is constant over the frequency range of interest in the vicinity of the resonant frequency. Using a commercial VNA, we measure the complex scattering parameters and evaluate the ratio of the reflected over the sent AC voltage $S_{11} = V_{\mathrm{out}}/V_{\mathrm{in}}$, also called the reflection coefficient Γ, which in terms of impedances can be expressed as [37],

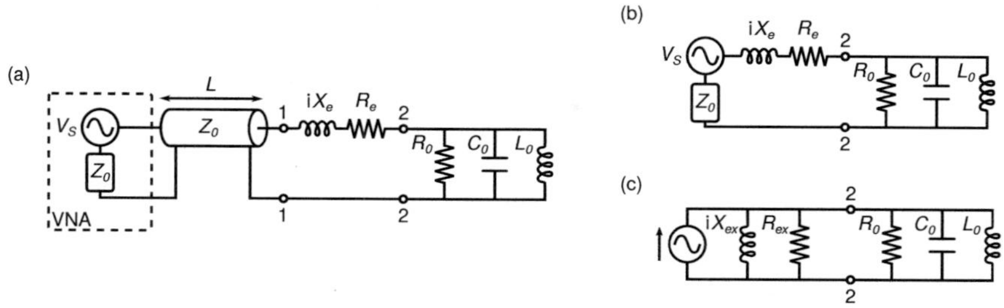

Figure 1.12: Electrical schematic of a parallel *RLC* resonator connected to external circuitry and the corresponding Thevenin and Norton equivalent circuits. **(a)** A resonator, modeled as a lumped element, parallel *RLC* circuit couples through an impedance ($R_e + iX_e$), and a transmission line with characteristic impedance Z_0 and length L to a vector network analyzer (VNA) with an impedance matched to Z_0. **(b)** Thevenin equivalent circuit of (a). Since the transmission line is terminated in a matched load, the input impedance seen from port 2 is independent of the length L of the transmission line [37]. The resonator with frequency $\omega_0 = 1/\sqrt{L_0 C_0}$ is therefore loaded by an external series impedance of $Z_0 + R_e + iX_e$. [85, 86].**(c)** Norton equivalent of the circuit shown in panel (b), where the Norton impedance parallel to the current source is split into a parallel combination of an external resistor R_{ex} and external reactance X_{ex}. From this circuit, one can define the coupling quality factor of the resonator as $Q_c = \omega_0 R_{ex} C_0$ and the coupling coefficient κ as the ratio between power dissipated in the resonator and power dissipated outside the resonator $\kappa = R_0/R_{ex}$.

$$S_{11} = \Gamma = \frac{Z_{in} - Z_{out}}{Z_{in} + Z_{out}} \tag{1.39}$$

From the equivalent circuit (cf. Fig. 1.12) it is now possible to calculate the frequency dependent reflection coefficient in terms of the resonator's quality factors [85, 86],

$$\Gamma(f) = \Gamma_d \frac{(1-\kappa)/(1+\kappa) + i2Q_l\delta(f)}{1 + i2Q_l\delta(f)}, \quad \text{with} \tag{1.40}$$

$$\delta(f) = (f - f_0)/f_0, \tag{1.41}$$

$$\kappa = Q_i/Q_l - 1. \tag{1.42}$$

Here, Γ_d accounts for a constant phase shift introduced by the coupling reactance X_e. Plotted in the real-imaginary plane, the reflection coefficient traces out a circle (cf. Fig. 1.13) which is the basis for a multitude of fitting algorithms, see for example Refs. [86–88].

In a real measurement two effects associated with the measurement setup and coupling can be observed. First any coupling reactance will rotate the entire circle away from the real axis. The same syndrom occurs if the reference plane of the VNA is not correctly set to take into account the signal delay caused by the length of transmission line. Secondly, any attenuation or amplification in the cryogenic measurement setup will rescale the diameter of the circle in the real-imaginary plane. This results in a shift of the so-called off resonant point of S_{11}, which is several linewidths detuned in frequency from the resonant frequency, from its expected value of $S_{11}^{\text{offres}} = 1 + i0$. To account for both effects, measured data is preprocessed such that the off resonant point lies at the expected value, which amounts to a rescaling of the amplitude signal, and the data is furthermore rotated such that the resonators circle is centered on the real axis. These operations largely take out the effect of the setup wiring from the measured data, which means that one can assume $\Gamma_d = 1 + i0$. For the fitting procedure we rely on the technique described in Ref. [86], which is crosschecked with an algorithm presented in Ref. [88][e].

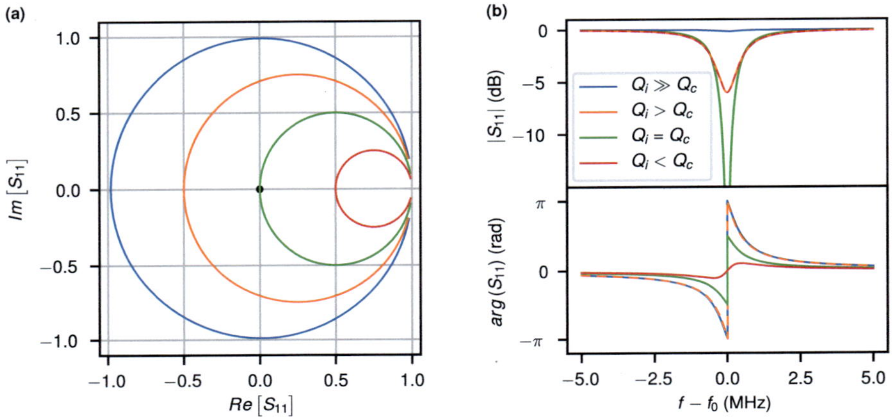

Figure 1.13: Complex reflection coefficient of a resonator measured in a single-port (reflection) setup. **(a)** Depending on the ratio Q_i/Q_c (see upper panel in (b) for the legend) three distinct regimes are visible: $Q_\text{i} > Q_\text{c}$ the circle has a diameter $d > 1$, which approaches $\lim_{Q_\text{i} \to \infty} d = 2$ (blue and orange circles), $Q_\text{i} = Q_\text{c}$ (green), the circle's diameter is exactly 1, and $Q_\text{i} < Q_\text{c}$ (red), where the circle has a diameter smaller than 1. **(b)** Amplitude (top) and phase (bottom) signal of the three regimes Q_i/Q_c. As can be seen, with lower Q_i the depth of the dip in the amplitude signal increases. However, the largest value is reached for a critically coupled resonator $Q_\text{i} = Q_\text{c}$. Checking the phase signal allows to distinguish whether $Q_\text{i}/Q_\text{c} > 1$, which leads to a 2π phase roll, or $Q_\text{i}/Q_\text{c} \leq 1$, which corresponds to a phase roll $\leq \pi$.

Figure 1.13(a) shows the complex reflection coefficient plotted in the real-imaginary plane. The circle can be clearly seen as well as the connection between its diameter d and the coupling coefficient κ of the resonator to the external circuitry, $\kappa = d/(2-d)$ [86]. Together with Eq. 1.40 three coupling regimes can be identified, see Table 1.5. Depending on whether most of the energy stored in the resonator is lost through coupling to the outside or by internal dissipation mechanisms, one operates in the *over-* or *undercoupled* regime. If the loss contributions from coupling and from internal dissipation mechanisms are exactly

[e] Python implementation avaliable at: https://github.com/sebastianprobst/resonator_tools

equal, the resonator is *critically coupled*. Since the diameter of the circle approaches $d \to 2$ for internal quality factors much larger than the coupling quality factor [cf. blue circle in Fig. 1.13(a)] and small differences in the calculation of the diameter lead to large changes in the coupling coefficient κ, a reliable measurement of the internal quality factors should be performed in a regime where Q_i is larger, but on the order of the coupling quality factor.

Table 1.5: The three coupling regimes for resonators connected to external circuits. Each regime comes with a characteristic reflection coefficient shape (cf. Fig. 1.13) whose hall mark signs are listed in the table below.

coupling regime	$Q_\mathrm{i}/Q_\mathrm{c}$	κ	d	amplitude $arg(S_{11})$
overcoupled	>1	>1	>1	2π
critically coupled	1	1	1	π
undercoupled	<1	<1	<1	$<\pi$

1.4.2 Dielectric loss

To quantify different loss mechanisms, the concept of participation ratios is used [8, 89–93]. It is based on the assumption that the total dielectric loss measured in a resonator $\tan(\delta_\mathrm{tot})$ is the sum of different loss channels with specific loss tangents $\tan(\delta_i)$, scaled by their corresponding participation ratio p_i. Since dielectric loss, which couples to the electric field of a resonator, is one of the main dissipation mechanisms in superconducting circuits, the participation ratio technique is mostly used to investigate which of the various interfaces contributes most to the total dielectric loss. Therefore, the participation ratio expresses the ratio of energy stored in the electric field in some specific volume V_i to the total energy stored in the resonator W_tot,

$$\tan(\delta_\mathrm{tot}) = \sum_i p_i \tan(\delta_i) \tag{1.43}$$

$$p_i = \frac{1}{W_\mathrm{tot}} \int_{V_i} d\vec{r}^3 \varepsilon_i |\vec{E}|^2. \tag{1.44}$$

In superconducting resonators the main contributions to the total dielectric loss originate in the following four areas: the metal-substrate interface, the metal-air interface, the substrate-air interface, and the bulk substrate [90–93]. Figure 1.14 schematically shows the location of different interfaces where dielectric loss in superconducting microwave resonators originates.

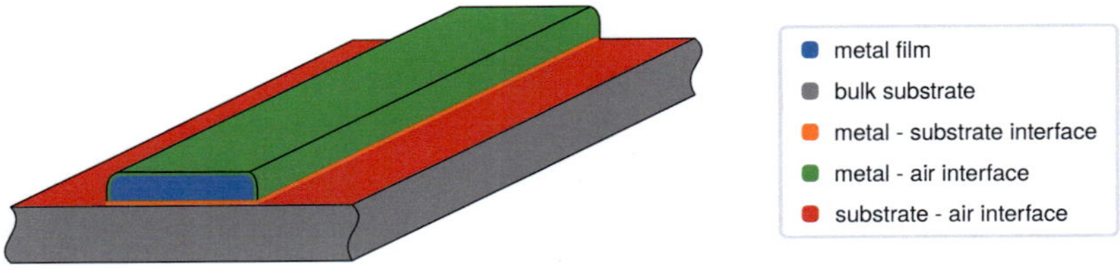

Figure 1.14: Schematic drawing of a microwave resonator fabricated from a thin metal film and the different interfaces responsible for dielectric loss.

Due to the fact that grAl is a granular material where pure metal grains are embedded in a matrix of non-stoichiometric AlO_x, we investigate dielectric loss as the limiting loss mechanism in grAl microwave resonators. Therefore, we compare the internal quality factors of different resonator geometries which cover a wide range of participation ratios. Since the resonators mainly investigated during the course of this thesis have a geometry similar to microstrip lines [cf. Fig 1.11(b)], albeit without the backside metalization usually providing the ground reference, for the investigation we concentrate on the substrate-air and metal-substrate interface participation ratios for our loss analysis, which are the main contributions for our geometry [90, 91]. We neglect the influence of the metal-air interface in the investigated geometries, since this interface has a participation ratio approximately one order of magnitude smaller than the other two, and a comparable loss tangent associated with it [92, 93]. This means the limit on Q_i imposed by this interface is significantly higher and therefore not taken into account.

Figure 1.15 shows the internal quality factors Q_i at an average number of photons circulating in the resonator $\bar{n} \approx 1$ ($\bar{n} = 4P_\mathrm{in}Q_\mathrm{l}^2 / \left(\hbar\omega_\mathrm{r}^2 Q_\mathrm{c}\right)$, with P_in the incident power on the sample holder [94]) for different sample geometries covering a surface participation ratio range of almost two orders of magnitude (for more details on the samples see Ch. 3). According to the results presented in Ref. [91] the surface loss tangent in superconducting circuits fabricated by lift-off technique can be well approximated by $\tan(\delta) = 2.6 \times 10^{-3}$ based on a comparison of published results from different research groups. Furthermore, recent results by Woods et al. [93] confirm the observed values of interface loss tangents. However, since in this work (Ref. [93]) resonators were fabricated by an etch process, where the metal-substrate interface remains pristine and is not covered with photoresist, the loss tangents for the metal-substrate interface are reported lower, which could be due to the fabrication process and cleaning steps involved.

As can be seen in Fig. 1.15 for higher surface participation ratios $p \gtrsim 10^{-3}$, the internal quality factors scale according to their participation ratio and are limited by a surface dielectric loss tangent $\tan(\delta) = 2.4 \times 10^{-3}$ in good agreement with previously reported results obtained on pure Al resonators. From these results we conclude that grAl microwave resonators are not limited by particularly lossy surface dielectrics. Decreasing the surface participation ratio further by using a 3D rectangular waveguide as a sample holder and no backside metalization (cf. Fig. 1.11(b) and Ch. 3) we find an internal quality factor on the order of $Q_\mathrm{i} = 1.5 \times 10^5$ comparable with previously reported superinductors from Josephson junction arrays or other disordered superconductors (cf. Sec. 1.3.1). These results already hint at sufficiently low microwave losses in granular aluminum thin films to realize superinductors in superconducting quantum circuits, where grAl should not limit the circuit coherence significantly below the state of the art.

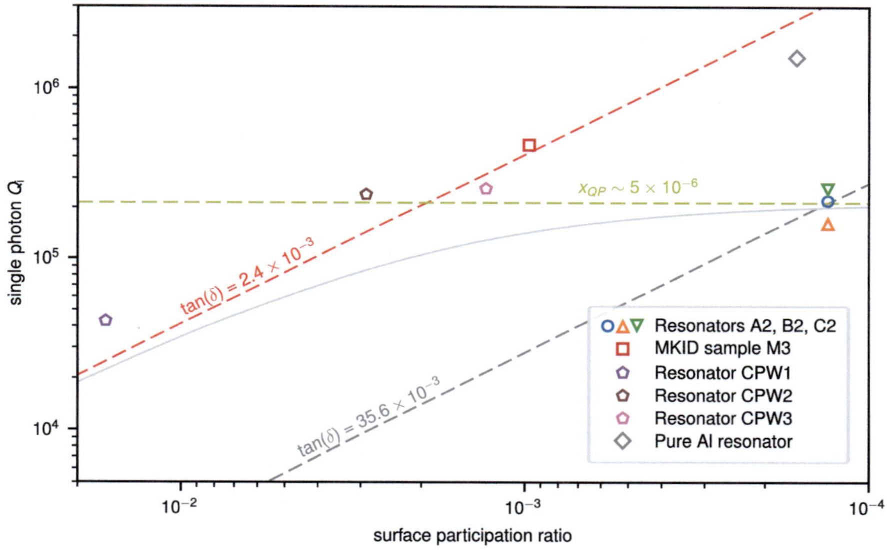

Figure 1.15: Internal quality factor Q_i in the single photon regime as a function of the surface participation ratio for different resonator geometries. For a surface participation ratio, which we define here as the sum the of substrate-air and metal-air interface participation ratios, $p \gtrsim 10^{-3}$ the measured Q_i of resonators fabricated from grAl scales according to $Q_\mathrm{i} \approx 1/(p\tan(\delta))$ with a loss tangent in good agreement with previously reported values in pure aluminum superconducting quantum circuits [91]. Granular aluminum resonators with a participation ratio $\sim 10^{-4}$ have an internal quality factor significantly lower than expected from dielectric loss and significantly lower than control samples in the same sample holder geometry but fabricated from pure aluminum [95]. From these results we conclude that grAl does not form particularly lossy dielectric interfaces and that it should be suited to be used in superconducting quantum circuits [46]. Furthermore, grAl resonators in a planar antenna geometry [cf. Fig. 1.11(b)] are likely limited by excess quasiparticles in the single photon regime at a level of $Q_\mathrm{i} \sim 1.5 \times 10^5$ (cf. Sec. 1.4.3).

A control experiment with resonators of similar participation ratio, also measured in the same 3D waveguide sample holder as grAl resonators with $p \sim 10^{-4}$, reveals internal quality factors as expected from the dielectric loss tangent. Therefore, we conclude that the antenna geometry grAl resonators are not limited by dielectric loss, but by excess quasiparticles as will be further discussed in Sec. 1.4.3.

1.4.3 Excess quasiparticles in grAl resonators

Following the discussion presented in Section 1.4.2, we find a strong indication for the fact that grAl resonators with a low surface participation ratio are not limited by dielectric loss. A large number of results presented in the literature show that the quasiparticle density in superconducting quantum circuits is several orders of magnitude higher than expected from the usual operating temperature $\sim 20\,\mathrm{mK}$ and the density of quasiparticles normalized to the total number of Cooper-pairs x_{QP}, the so called excess quasiparticle density is on the order of 10^{-6} [16, 39–44]. If one assumes that the loss limiting the internal quality factor of grAl resonators A2, B2, and C2 (cf. Fig. 1.15) is entirely dominated by excess quasiparticles, we calculate an excess quasiparticle density $x_{\mathrm{QP}} = 1/(p_{\mathrm{QP}}Q_{\mathrm{i}}) \approx 5 \times 10^{-6}$, which is in the range of previously reported values.

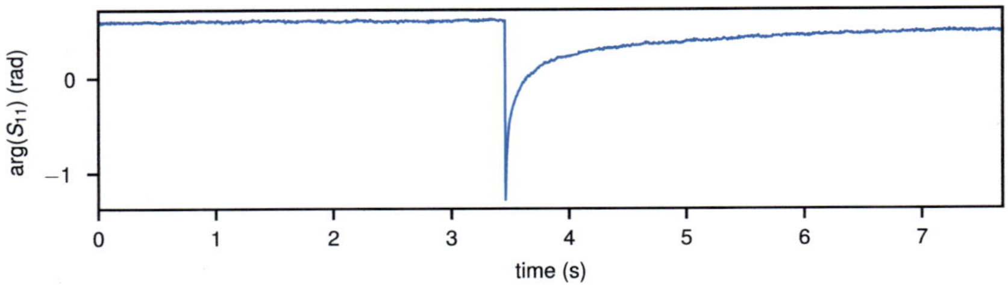

Figure 1.16: Measurement of a quasiparticle producing event and its subsequent relaxation. After a sudden drop of the resonant frequency, indicated by a lower $arg(S_{11})$, the resonator relaxes back to its equilibrium state over a period on the order of seconds. This time scale is orders of magnitude longer than previously observed in KIDs fabricated from different materials.

Through a continuous monitoring of the phase signal of quasiparticle limited grAl resonators, and subsequent inversion of the resonator's phase response as a function of frequency, we observe sudden drops in the resonant frequency followed by a relaxation back to the equilibrium value over a timescale of seconds (see Fig. 1.16) [46]. The signature of the observed events is reminiscent of the signal shape produced by the impact of high energy particles in kinetic inductance detectors (KIDs) [96–98]. Breaking a large number of Cooper pairs by energy deposition into the resonator increases the kinetic inductance of the superconducting wire due to the fact that $L_{\mathrm{kin}} \propto 1/n_{\mathrm{CP}}$. The resulting frequency shift of the resonator is monitored and constitutes the signal in KIDs, which are used as sensitive detectors in radio astronomy [99, 100]. Figure 1.16 shows the signature of such a quasiparticle producing event measured in a resonator fabricated from a 20 nm thick grAl film with resistivity $\rho = 4 \times 10^3\,\mu\Omega\,\mathrm{cm}$. The sudden drop of the phase signal, corresponding to a drop of the resonant frequency, at a time scale much faster than resolvable with VNA is clearly visible. In addition, the plot shows the unexpectedly long time on the order of seconds over which the resonant frequency relaxes back to the equilibrium value. This time scale is several orders of magnitude longer than what has been observed previously in KIDs made from pure aluminum [96].

By performing a long measurement over 45 min on two of the grAl resonators shown in Fig. 1.11(b), we find an average event rate of $\Gamma_b = 1/20\,\mathrm{s}$ (see Fig. 1.17. Furthermore, by performing the same measurement shown in Fig. 1.17 at different readout powers, i.e. different average numbers of photons circulating in the resonator, we find a faster relaxation time after the impacts with an increased photon number. As the internal quality factor of the resonators also increases with higher photon numbers, we developed a phenomenological model attributing the increase in Q_i to a faster quasiparticle recombination and a resulting smaller excess quasiparticle density (for details see Sec. 3.4).

The results briefly summarized in this section classify granular aluminum as a potential material for the fabrication of superinductors in superconducting quantum circuits. At the same time, they indicate that high energy impacts could be a source of the unexpectedly large excess quasiparticle density observed in several quantum circuits fabricated from pure aluminum, and show the need for further investigation into the origin of the impacts and possible strategies to mitigate their effects in order to allow longer coherence times of superconducting qubits. For a more detailed discussion of the presented summary, we refer the reader to Ch. 3.

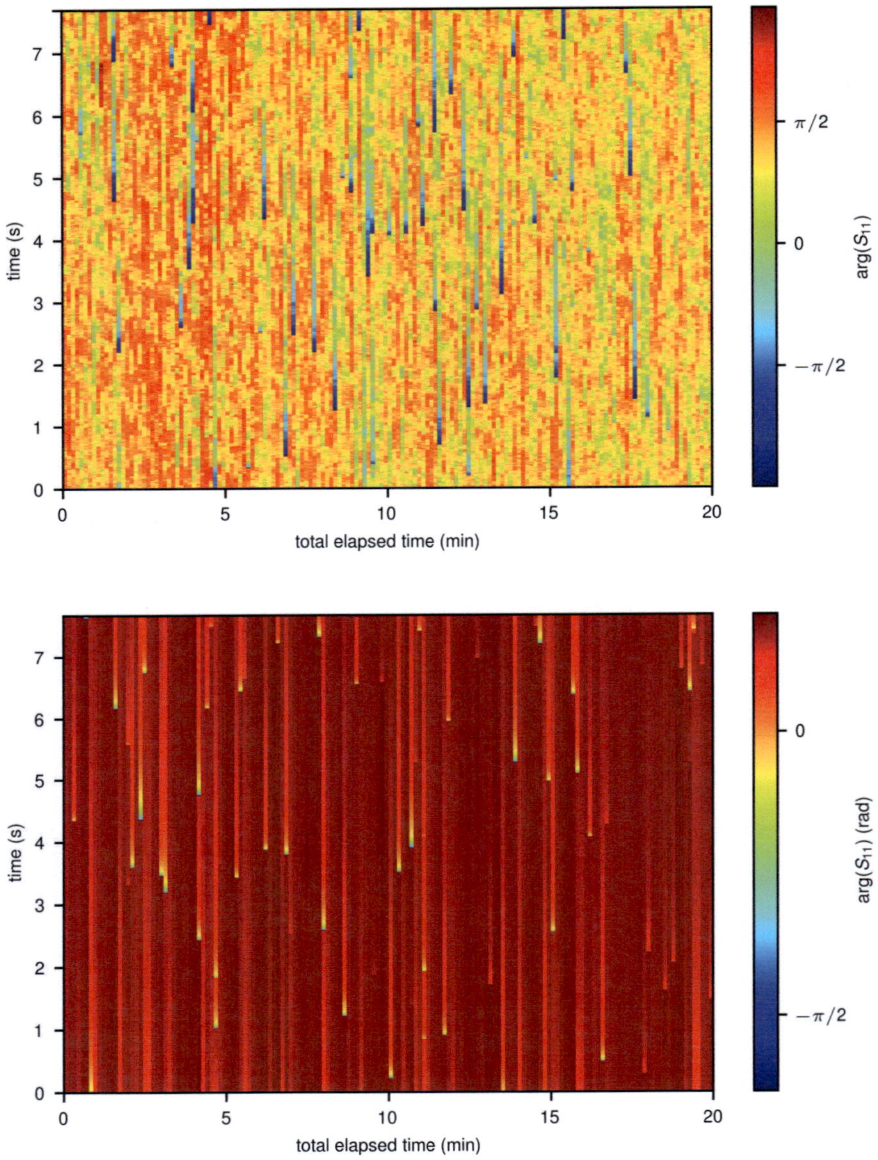

Figure 1.17: Measurement of the rate of quasiparticle producing events. Measurements at readout powers corresponding to $\sim 10^2$ average photons circulating in the resonator of the weakest coupled antenna geometry grAl resonator ($Q_c = 60 \times 10^3$, top panel), and strongest coupled resonator ($Q_c = 4 \times 10^3$, bottom panel) are shown. The smaller linewidth of the weaker coupled resonator results in a higher sensitivity to QP producing events. From a statistical evaluation of the time between events we find an exponential distribution of the time, indicative of a Poisson process of individual, uncorrelated stochastic events. We extract an average rate of the bursts $\Gamma_b = 1/20/\mathrm{s}$. For a more detailed discussion see Ch. 3.

1.5 Fluxonium qubit case study

Measurements of microwave resonators fabricated from grAl show encouraging results to use the material for the fabrication of superinductors in superconducting quantum circuits. As discussed in Section 1.3, granular aluminum offers a large, tunable kinetic inductance, and microwave losses comparable to Josephson junction arrays, while maintaining much smaller nonlinearities similar to other disordered superconductors. As a case study for the performance of grAl superinductors, we designed, fabricated, and measured a superconducting fluxonium qubit. The present section briefly summarizes the obtained results. A more thorough discussion of the design, fabrication, measurements, and results can be found in Ch. 4.

1.5.1 Design

To briefly recall the discussion presented in Section 1.1, the superconducting fluxonium qubit is a small Josephson junction with an area $A \sim 0.1\,\mu\text{m}^2$, a Josephson energy $E_\text{J}/h \sim 10\,\text{GHz}$ and a junction plasma frequency $\omega_\text{p}/2\pi \sim 20\,\text{GHz}$ shunted by a large, linear inductor with an inductance $L \sim 10^2\,\text{nH}$, with an impedance $Z \geq R_\text{Q}$, and lowest self-resonant mode of the inductor sufficiently above the qubit frequency operating range, i.e. $\gtrsim 12\,\text{GHz}$.

To read out the qubit state, we use the tools of cQED (cf. Sec. 1.1): by coupling the fluxonium qubit via a shared inductor to a readout resonator, we can measure the shift of the readout resonator's frequency, which depends on the state of the qubit. The top panel in Figure 1.18 shows an optical microscope image of a fluxonium (yellow rectangle) coupled to a readout resonator. Via its dipole moment, the resonator couples to a 3D waveguide sample holder (cf. Ch. 4 for more experimental details), which connects to the cryogenic measurement setup (cf. Appendix C.1). The two large pads at both ends of the antenna allow to measure the room temperature resistance of the grAl wires, which implement the resonator inductance L_r and the shared inductance L_s. From the measured resistance we can estimate the kinetic inductance via the Mattis-Bardeen formula (cf. Sec. 1.3), i.e. the pads allow a quick pre-characterization of the fabricated device.

As can be seen in the false colored scanning electron microscope (SEM) image in the center panel of Fig. 1.18, all inductive elements besides the small Josephson junction are implemented with granular aluminum wires (false colored in red). The Josephson junction, which provides the nonlinearity to the circuit is a Al/AlO_x/Al Josephson junction fabricated in a hybrid Niemeyer-Dolan/bridge-free technique (see. Sec. 4.1.3 and Appendix A) [59–61]. All inductors of the circuit, which are the readout resonator inductor L_r, the shared inductor between resonator and fluxonium L_s, which provides the coupling, and the superinductor L_q can be independently tuned by adjusting the length of the corresponding wires. Comparing the fabricated design shown in Fig. 1.18 with a fluxonium qubit with a Josephson junction superinductor, for example shown in Ref. [101], or a NbN nanowire superinductor presented in Ref. [26], illustrates the reduced design complexity enabled by the use of granular aluminum.

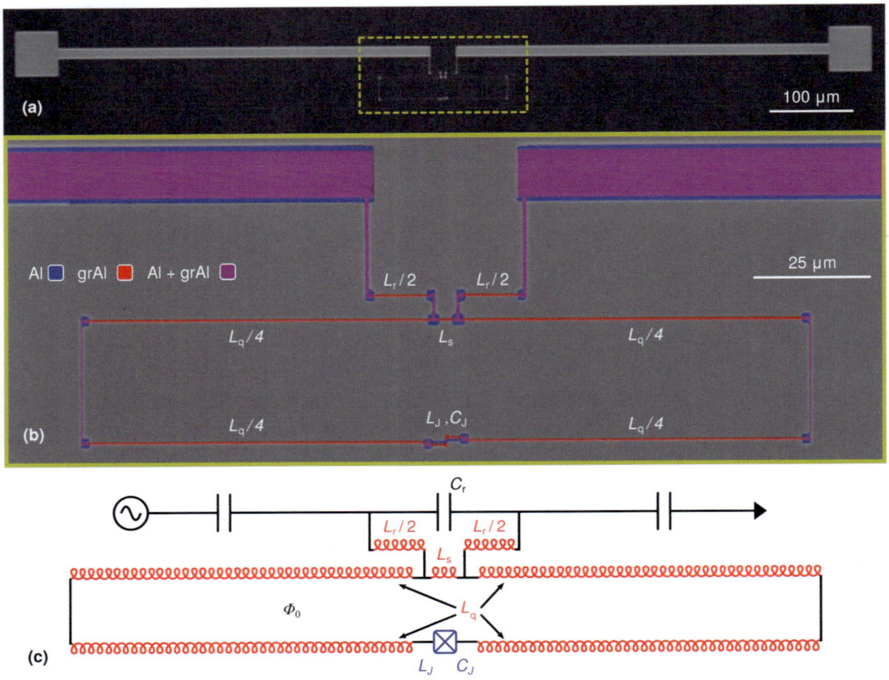

Figure 1.18: Superconducting fluxonium qubit with a grAl superinductor. **(a)** Optical microscope image of the readout resonator realized as an antenna coupled to a fluxonium qubit. The resonator couples via its dipole moment to the TE_{10} mode of a rectangular waveguide, which serves as the sample holder and connection to the cryogenic measurement setup following Ref. [102]. **(b)** False colored optical microscope image showing the resonator-qubit coupling region and the fluxonium loop [cf. yellow rectangle in panel (a)]. All inductive elements of the circuit, the resonator inductor L_r, the shared inductor L_s mediating the coupling between resonator and fluxonium, and the superinductor L_q are realized as grAl wires. The independent control of their lengths enables the adjustment of the corresponding inductances. **(c)** Electrical schematic of the fluxonium inductively coupled to a readout resonator. The circuit elements are colored according to the material they are fabricated from [see legend in panel (b)].

1.5.2 Spectroscopy

Using the dispersive readout technique of cQED [49, 50], we perform standard two-tone qubit spectroscopy. By monitoring the amplitude and phase response of the readout resonator with a commercial vector network analyzer (VNA), while sweeping a continuous wave microwave tone in frequency, we measure the fluxonium spectrum as a function of the externally applied flux Φ_{ext}, see Fig. 1.19. Through a coil placed around the waveguide sample holder, we apply the magnetic field controlling the external flux (cf. Appendix C.1).

Figure 1.19 shows the measured fundamental $|g\rangle - |e\rangle$ transition (blue points) together with a numerical fit of the fluxonium Hamiltonian to the measured data (solid line). The plot illustrates the large frequency tuning range of the qubit transition frequency and the excellent agreement between the measured data and the theoretical model. Based on the agreement between fit and data we conclude that the grAl wires shunting the small Al/AlO$_x$/Al Josephson junction behave as a superinductor in the previously defined sense: it does not introduce additional nonlinearity or spurious resonant modes below 12.5 GHz into the circuit.

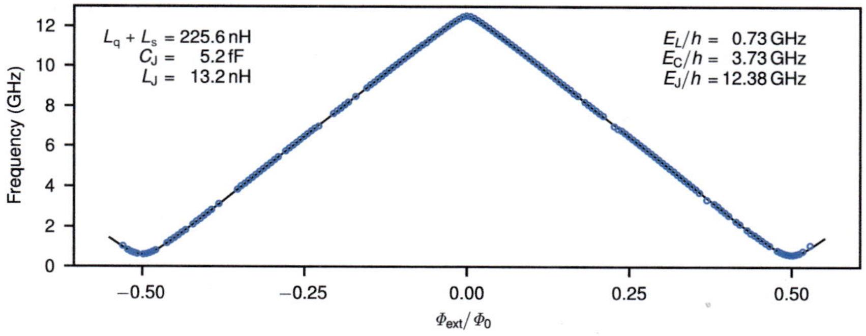

Figure 1.19: Granular aluminum fluxonium spectrum. Employing a standard two-tone microwave measurement we extract the fundamental grAl fluxonium transition (blue points). Numerically fitting the fluxonium Hamiltonian to the measured data (black line) yields the circuit parameters, which are given in the plot. By changing the external magnetic flux, we can tune the fluxonium qubit frequency from 595 MHz to 12.5 GHz. From the good agreement between the numerical fit and the measured data we conclude that the grAl wire behaves as a linear, lumped-element superinductor.

From the numerical fit, we extract a superinductor with $L_q + L_s = 225.6\,\text{nH}$, which corresponds to an inductive energy $E_L/h = (\hbar/2e)^2/(L_q + L_s) = 0.73\,\text{GHz}$. From the fitted value for the total capacitance of the circuit $C_t = 5.2\,\text{fF}$, we attribute 3 fF to the Josephson junction based on a typical Al/AlO$_x$/Al Josephson junction capacitance of 50 fF/µm^2 and a junction area of $A \approx 0.06\,\mu\text{m}^2$ measured using a scanning electron microscope image. Following this argument, the superinductor loop has a capacitance of $C_q = 2.2\,\text{nH}$, which leads to a characteristic impedance of $Z = \sqrt{(L_q + L_s)/C_q} \approx 10\,\text{k}\Omega$, well above the resistance quantum

$R_Q = 6.5\,\mathrm{k}\Omega$. The small Josephson junction, providing the circuits nonlinearity adds an inductance $L_J = 13.2\,\mathrm{nH}$, corresponding to a Josephson energy $E_J/h = 12.38\,\mathrm{GHz}$. With a ratio of $E_J/E_C = 3.3$, with $E_C = e^2/2C_t$, our fluxonium is in the range of typical fluxonium parameters [15–18]. At the half integer sweet spot $\Phi_{ext}/\Phi_0 = n + 0.5$, the qubit frequency goes as low as $f^q_{0.5} = 0.594\,\mathrm{GHz}$, which can be tuned up to $f^q_0 = 12.538\,\mathrm{GHz}$ at the integer flux sweet spot $\Phi_{ext}/\Phi_0 = n$, with $n \in \mathbb{Z}$ by the externally applied magnetic flux.

1.5.3 Coherence

Time resolved, and pulsed spectroscopy provide the final benchmarking of granular aluminum superinductors for quantum circuits. By performing free decay energy relaxation measurements to obtain T_1, and Ramsey fringes measurements which yield T_2 [103], we are able to assess the fluxonium coherence and can estimate its limiting mechanisms. These results should provide an answer to the question about the suitability of granular aluminum for its application as a material for superinductors in high coherence quantum circuits. Although Sections 1.3 and 1.4 already provide indications about the performance of the material, the fluxonium case study presented here should give an answer for the suitability of grAl when implemented in actual superconducting quantum circuit devices.

The fluxonium spectrum possesses two families of operating points with favorable properties for qubit coherence, the so-called sweet spots (cf. Fig. 1.19). One of them, which is when the externally applied magnetic field is such that the flux threading the fluxonium loop $\Phi_{ext}/\Phi_0 = n + 0.5$, with $n \in \mathbb{Z}$ is the standard operation point for the fluxonium qubit, and referred to as a *half flux* sweet spot. Regarding the energy relaxation time T_1, the dissipation due to quasiparticles tunneling through the Josephson junction barrier vanishes if the phase drop across the Josephson junction is exactly π [19]. Since this dissipation mechanism is suppressed, T_1 can exhibit a local maximum, if the limiting loss mechanism is indeed quasiparticle tunneling through the junction.

The second reason for the half-flux operation point of the fluxonium qubit is the fact that the derivative of the transition frequency with respect to the external magentic flux vanishes. Therefore, the qubit frequency is to first order insensitive to small fluctuations of the external magnetic field. Similar to the flux qubit or tunable transmon qubits, this leads to a sweet spot for the phase coherence of the fluxonium T_2, which ultimately is the limiting timescale for any qubit in the context of quantum information processing.

The second sweet spot for the operation of the fluxonium qubit ist the so-called *zero flux* sweet spot. Here, the qubit is susceptible to dissipation by quasparticle tunneling over its small junction, but the spectrum exhibits a first order flux noise insensitive point, which is favorable for T_2.

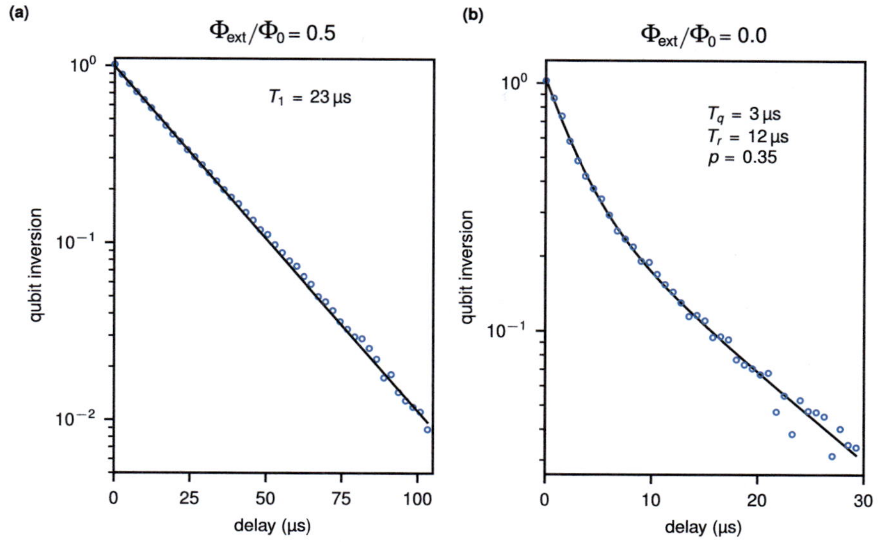

Figure 1.20: Energy relaxation time T_1 at both fluxonium sweet spots. **(a)** At the $\Phi_{\text{ext}}/\Phi_0 = 0.5$ sweet spot, where dissipation due to quasiparticle tunneling across the Josephson junction is suppressed (cf. Sec. 1.2.3), we measure a single exponential decay (blue points). From the exponential fit (black line) we extract an energy relaxation time $T_1 = 23\,\mu\text{s}$. **(b)** At the $\Phi_{\text{ext}}/\Phi_0 = 0.0$ sweet spot, we observe a double exponential decay, which is indicative of dissipation caused by quasiparticle tunneling [16, 43, 104]. From the fit of a sum of two exponentials to the measured data (blue points) we extract a decay due to quasiparticle dissipation $T_q = 3\,\mu\text{s}$, a residual relaxation rate $T_r = 12\,\mu\text{s}$, and a ratio of $p = 0.35$ between decays limited by T_q and T_r.

Figure 1.20 shows the measured energy relaxation time T_1 of the fluxonium qubit at both sweet spots. Panel (a) is the measurement at $\Phi_{\text{ext}}/\Phi_0 = 0.5$, where the fluxonium is not susceptible to dissipation by quasiparticles tunneling across the Josephson junction. Here, we measure an average T_1 time of 23 µs comparable to the current state of the art in superconducting qubits. Besides quasiparticle tunneling across the small Junction, which is suppressed at the half-flux sweet spot, excess quasiparticles in the superinductor or dielectric loss could very well be the main loss mechanisms. Assuming the limiting loss mechanism for the measured T_1 of 23 µs is inductive loss in the superinductor, following the methodology of Refs. [16, 34, 35] (cf. Tab. 1.2 in Sec. 1.2.3), we estimate an excess quasiparticle density normalized to the number of Cooper-pairs,

$$x_{\text{QP}}^{\text{SupInd}} = 4.5 \times 10^{-7}. \tag{1.45}$$

The extracted value of $x_{\text{QP}} = 4.5 \times 10^{-7}$ is about one orders of magnitude larger than previously observed in a fluxonium qubit with a Josephson junction array superinductor [16], which could be due to the long quasiparticle relaxation time observed in granular aluminum (cf. Sec. 1.4.3).

Since capacitive loss $\propto \omega_{01}^2$ (cf. Sec. 1.2.3), a T_1 at $\Phi_{\mathrm{ext}}/\Phi_0 = 0.5$ limited by lossy dielectrics in the fluxonium capacitances, where the fluxonium qubit frequency reaches its minimum would imply a T_1 at $\Phi_{\mathrm{ext}}/\Phi_0 = 0.0$ about a factor of ten smaller than observed in our experiments. For a more detailed discussion see Sec. 4.3.1.

At the zero flux sweet spot, we observe a double-exponential decay as shown in Fig. 1.20(b), which is indicative of relaxation dominated by the dissipative tunneling of excess quasiparticles across the small Josephson junction of the fluxonium qubit [16, 43, 104]. Therefore, we assume that the limiting loss mechanism for T_q is indeed the excess quasiparticle density close to the Josephson junction, which can be estimated as described in Sec. 1.2.3,

$$x_{\mathrm{QP}}^{\mathrm{JJ}} = 1.5 \times 10^{-5}. \tag{1.46}$$

This value is calculated using a superconducting energy gap of thin film aluminum $\Delta_0 = 210\,\mu\mathrm{eV}$, and Josephson energy $E_\mathrm{J}/h = 12.38\,\mathrm{GHz}$, which we obtained from the numerical fit to the fluxonium spectrum (see. Sec. 1.5.2). The extracted excess quasiparticle density $x_{\mathrm{QP}}^{\mathrm{JJ}}$, is about an order of magnitude larger than previously reported in superconducting circuits fabricated from pure aluminum thin films [16, 39–44]. A possible explanation for this fact could be the larger superconducting gap of the grAl forming the superinductor $\Delta_0 = 300\,\mu\mathrm{eV}$ compared to thin film aluminum $\Delta_0 = 210\,\mu\mathrm{eV}$. Therefore, quasiparticles could potentially be trapped in the lower gap region of the Josephson junction vicinity and increase the effective excess quasiparticle density.

Figure 1.21(a) shows a Ramsey fringes measurement [103] at the $\Phi_{\mathrm{ext}}/\Phi_0 = -0.5$ sweet spot (black points). From a fit of an exponentially decaying cosine to the measured data we extract a fluxonium qubit coherence time $T_2 = 28\,\mu\mathrm{s}$, which is comparable to coherence times of Josephson junction array superinductor fluxoniums [16, 17, 105]. By inserting a π-pulse in the middle of the Ramsey sequence, a so-called spin-echo measurement, we partly filter out low frequency noise and observe an increase in coherence up to $T_2^{\mathrm{E}} = 46\,\mu\mathrm{s}$, close to the maximum theoretically possible value $T_2^{\max} = 2T_1$. Panel (b) of Fig. 1.21 shows the dependence of T_2 as a function of the externally applied flux. As can be seen, for values slightly away from the first-order flux noise insensitive point of the fluxonium spectrum the coherence time T_2 rapidly decreases as the qubit frequency becomes more and more susceptible to the externally applied flux. In between the mentioned half flux sweet spots and the integer flux sweet spots the frequency of the fluxonium qubit far reaches its maximum sensitivity to the externally applied flux $\partial f_{01}/\partial\Phi_{\mathrm{ext}} \sim 24\,\mathrm{GHz}/\Phi_0$ (cf. Fig. 1.19). Consequently, T_2 is reduced to values on the order of tens of nanoseconds. For a more detailed discussion of the coherence properties of the granular aluminum fluxonium see Ch. 4.

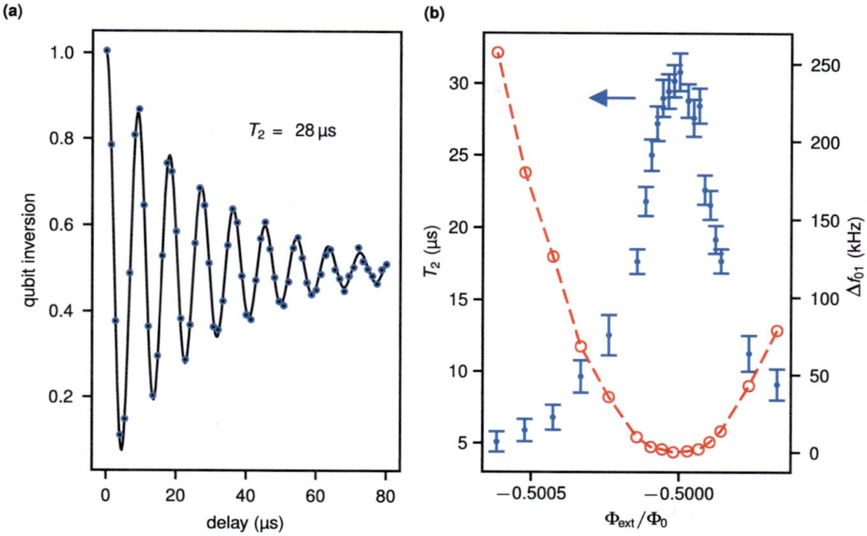

Figure 1.21: Coherence of the fluxonium qubit at the $\Phi_{ext}/\Phi_0 = -0.5$ sweet spot. **(a)** By fitting an exponentially damped cosine (black line) to the measurement data of a Ramsey fringes measurement, we extract a coherence time $T_2 = 28\,\mu s$. **(b)** T_2 as a function of the external applied magnetic flux. Due to the significantly increasing susceptibility to flux noise in the fluxonium, only $5\,\mu\Phi_0$ away from the sweet spot, T_2 is decreased by a factor ~ 6. In the same flux interval the fluxonium frequency merely changes by $\Delta f_{01} = f_{01}(\Phi_{ext}) - f_{01}(0.5\Phi_0) \approx 250\,kHz$.

Based on the results presented in this introductory section, we conclude that granular aluminum is a suitable option to implement superinductors in superconducting quantum circuits. The in-situ integrability with standard $Al/AlO_x/Al$ Josephson junction fabrication and relatively straightforward deposition technique, should enable the design and realization of increasingly complex quantum circuits. A main task to allow further advancement in circuits using granular aluminum is the understanding and mitigation of the dominating loss and decoherence mechanisms, which we will discuss further in Chapter 4.

2 Granular aluminum in the literature

In the following chapter we will attempt to give a short overview of the research carried out towards understanding the dome-shaped behavior of the critical temperature T_c of grAl as a function of its normal state resistivity. In the nearly 60 years since the initial discovery of the enhanced T_c in grAl, its normal state transport properties have also been the subject of research, which we will not go into here. For a broader and more detailed discussion of the multi-faceted research involving granular aluminum, we refer the reader to Ref. [106], and recent publications [107–111].

2.1 Overview of experimental results

In the following section we discuss experimental results regarding the superconducting properties of granular aluminum, which is still an active area of research. Fig. 2.1 shows recent results of experiments investigating the critical temperature T_c and superconducting gap Δ_0 of granular aluminum as a function of its normal state resistivity [82]. In their experiments, the authors harnessed the response of microwave kinetic inductance detectors to extract the relevant energy scales governing superconductivity. Figure 2.1 serves as a reference point for the following discussion, to connect resistivity, superconducting gap and critical temperature.

More than 60 years before the results in Fig. 2.1 were obtained, Buckel and Hilsch [113] first measured an enhanced critical temperature compared to the bulk critical temperature of 1.2 K in aluminum thin films. Following the evaporation of pure Al on a quartz substrate at 4 K, they observed a $T_c = 2.59\,\mathrm{K}$ in a 20 nm thick Al film. To measure the superconducting transition temperature the authors perform a four probe measurement to monitor the resistivity of $10\,\mathrm{mm} \times 1\,\mathrm{mm}$ disordered aluminum strips as a function of temperature. Interestingly, after thermal cycling their sample to room temperature the enhanced T_c vanished. In their study Buckel and Hilsch attribute this to recrystallization of the aluminum film resulting in the removal of lattice defects.

About ten years after the results by Buckel and Hilsch the interest in granular superconductors and their enhanced T_c increased. In experiments by Abeles *et al.* in 1966 [114] the authors observed an increased T_c together with a decreased grain size in aluminum films deposited in a slight oxygen atmosphere. They measured a maximum $T_c = 2.3\,\mathrm{K}$ in films deposited on a substrate at room temperature by employing DC resistance measurements. Using a substrate at 100 K they reduced the aluminum grain size even further and measured critical temperatures up to $\sim 3\,\mathrm{K}$.

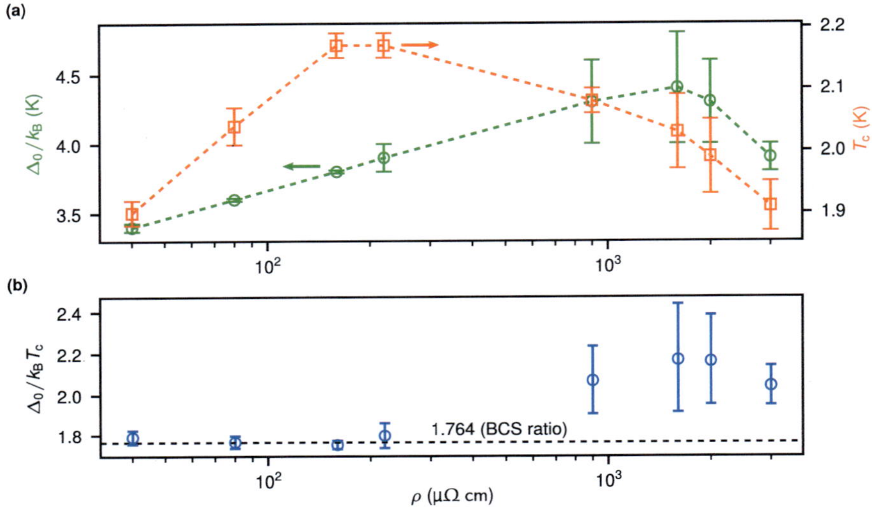

Figure 2.1: Superconducting gap Δ_0, critical temperature T_c and ratio $\Delta_0/k_B T_c$ as a function of the normal state resistivity ρ for granular aluminum films deposited on a substrate at room temperature (see Ref. [82]). **(a)** Superconducting gap Δ_0 (green, left axis) and critical temperature (orange, right axis) as a function of ρ. With larger resistivity both Δ_0 and T_c increase. At $\rho \approx 2 \times 10^2\,\mu\Omega\,\mathrm{cm}$ the critical temperature reaches a maximum and subsequently decreases again, while the highest values for Δ_0 are observed at $\rho \approx 2 \times 10^3\,\mu\Omega\,\mathrm{cm}$. **(b)** Ratio between superconducting gap and critical temperature. For lower resistivities, where both Δ_0 and T_c increase as a function of ρ we observe ratios of $\Delta_0/k_B T_c$ close to the BCS ratio of 1.764 [112]. At higher resistivities where the critical temperature decreases, while the gap is still enhanced, we observe ratios significantly larger than the standard BCS ratio.

In 1968, summarizing their previous results, Cohen and Abeles [115] presented experimental data on the critical temperature, critical magnetic field, critical current, and energy gap of granular aluminum films. Due to the addition of oxygen, aluminum films presumably did not recrystallize, which stabilized the enhanced T_c in contrast to the earlier results by Buckel and Hilsch [113]. Furthermore, Cohen and Abeles suggested that the oxygen likely precipitates in the form of oxide at the grain boundaries. It was also then that they coined the term *granular superconductor* for a material system of superconducting grains separated by tunneling barriers, which was previously introduced by Parmenter [116]. A surprising result was the sharp superconducting gap of granular aluminum measured via the conductance of Al/AlOx/Pb and Al/AlOx/Sn tunnel junctions. Based on the assumption of an inhomogeneous granular structure of grAl Cohen and Abeles expected to find a distribution of Δ_0, which was not the case.

Deutscher *et al.* [83] provided further insight into the microstructure of granular aluminum films and how grain size, normal state resistivity, and superconducting transition temperature are related. By employing electron diffraction, they showed in 1973 that in all their investigated samples, which covered resistivities from $\rho = 3\,\mu\Omega\,\mathrm{cm}$ to $1 \times 10^6\,\mu\Omega\,\mathrm{cm}$, the lattice spacing was equal to that of bulk aluminum. This means that even in highly resistive grAl films the grains consist of pure, crystalline aluminum.

A new experimental approach of using dark-field electron microscopy and measuring the grain sizes of about 10^3 grains in every sample allowed the authors of Ref. [83] to obtain grain size histograms. Based on these results they distinguished two classes of samples, which separate at a resistivity $\rho = 10^2\,\mu\Omega\,\mathrm{cm}$. For samples below this resistivity, increased oxygen pressure during film deposition leads to decreasing grain size d. For films deposited on substrates at room temperature d eventually saturates at $\sim 3\,\mathrm{nm}$. Furthermore, with higher resistivity the grain size distribution narrows and the tail of large grain sizes observed in low resistivity films decreases. In this first resistivity regime the critical temperature measured by DC transport experiments increases approximately proportional to d^{-1} indicating a size effect as the underlying mechanism for the enhancement of T_c in granular aluminum.

In samples with resistivities $\rho > 10^2\,\mu\Omega\,\mathrm{cm}$ the grain size essentially remains constant at $d \sim (3 \pm 1)\,\mathrm{nm}$, but insulating barriers between grains build up. In granular aluminum films with resistivities larger than $5 \times 10^2\,\mu\Omega\,\mathrm{cm}$, T_c decreases and the authors of Ref. [83] observed increasing widths of the transition to the superconducting phase. For grAl films with $\rho \geq 10^7\,\mu\Omega\,\mathrm{cm}$ only incomplete transitions to the superconducting state were observed.

Many recent investigations of granular aluminum employ THz spectroscopy in addition to DC transport measurements [84, 111, 117, 118]. These measurements are typically performed in a frequency range of about 100 GHz to 700 GHz and allow to extract the complex transmission coefficient of granular aluminum films. Fitting the Mattis-Bardeen theory [119] to the measured complex conductivity yields values for the superconducting gap Δ_0 and the superfluid stiffness J [84, 120]. However, similar to DC transport measurements, THz spectroscopy only gives access to the properties of the bulk samples, and does not permit to investigate the superconducting properties of individual grains.

Spectroscopic measurements on kinetic inductance detectors (KIDs) in a frequency range up to 300 GHz with a resolution of 1 GHz [82] also give access to the superconducting properties of grAl averaged over a KID, but the increased resolution in comparison to previous experimental results allows to resolve features below the superconducting gap. The signal measured in Ref. [82] by monitoring the phase and amplitude response of the fundamental mode of their KIDs is the combined effect of Cooper-pair breaking in the grAl film, and the excitation of higher modes in the microwave resonators that are used as KIDs. The breaking of Cooper-pairs decreases the internal quality factor of superconducting microwave resonators (cf. Sec. 1.4) and lowers their resonant frequency via an increase in the kinetic inductance proportional to the number of Cooper-pairs. Additionally, due to the nonlinearity of the kinetic inductance in grAl (cf. Sec. 1.3.2) the excitation of higher modes in KID resonators results in a frequency shift of the monitored lowest mode via the cross-Kerr interaction. A combination of the previously described THz transmission spectroscopy on grAl films and spectroscopy in KIDs fabricated from grAl might allow to further investigate the nature of the observed sub gap modes in granular aluminum [82, 84, 117, 118].

To summarize the presented experimental results, the following is known about the superconducting properties of granular aluminum:

- Up to a normal state resistivity $\sim 5\times10^2\,\mu\Omega\,\mathrm{cm}$ the size of the pure aluminum grains d in the non-stoichiometric AlOx matrix decreases as a function of the resistivity ρ [83]. It saturates at $d=(3\pm1)\,\mathrm{nm}$ for a deposition on substrates at room temperature, while it goes down to $d\lesssim 2\,\mathrm{nm}$ for a deposition on substrates at $T\leq 100\,\mathrm{K}$ [121]. In films with $\rho\gtrsim 5\times10^2\,\mu\Omega\,\mathrm{cm}$ d of the pure aluminum grains does not change, but they progressively decouple because of a build up of the insulating barriers between them.
- The superconducting energy gap Δ_0 as measured by THz spectroscopy increases up to a resistivity of $\rho\sim 10^3\,\mu\Omega\,\mathrm{cm}$, plateaus and decreases again for resistivities $\rho\gtrsim 3\times10^3\,\mu\Omega\,\mathrm{cm}$ [82, 84], see Fig. 2.1(a).
- Granular aluminum films deposited on a substrate at $T\leq 100\,\mathrm{K}$ show a dome shaped T_c as a function of the resistivity. At $\rho\sim 3\times10^2\,\mu\Omega\,\mathrm{cm}$ T_c reaches its maximum of $\sim 3.15\,\mathrm{K}$ and subsequently decreases [84, 107, 108, 111, 122].
- Granular aluminum films deposited on a substrate at room temperature also show a dome shaped T_c with resistivity. However, the maximum T_c is $\sim 2.17\,\mathrm{K}$ for a normal state resistivity $\rho\sim 2\times10^2\,\mu\Omega\,\mathrm{cm}$ [82, 83].
- The superfluid stiffness J decreases approximately proportional to $1/\rho$ in granular aluminum films regardless of the substrate temperature during film deposition [82, 84].
- The ratio $\Delta_0/(k_B T_c)$ increases from the standard BCS value 1.764 for granular aluminum films with resistivities $\rho\lesssim 10^3\,\mu\Omega\,\mathrm{cm}$ to a ratio of 2.1 to 2.2 for films with higher resistivities [82, 84, 111], see Fig. 2.1(b).
- Granular aluminum films deposited on room temperature substrates undergo a superconductor to insulator transition at resistivities $\rho > 10^4\,\mu\Omega\,\mathrm{cm}$ [82], whereas for

grAl films deposited on a cold substrate the transition occurs at higher resistivities $\rho \sim 10^5\,\mu\Omega\,\mathrm{cm}$ [84, 111, 122].

2.2 Possible origins of enhanced critical temperature

In the following section we give an overview over theoretical proposals to explain the enhanced critical temperature in granular aluminum. We will refrain from detailed discussions of the underlying theoretical calculations and refer the expert reader to the correspondingly cited literature for deeper theoretical insight.

As already pointed out in the previous section the size of the pure aluminum grains in granular aluminum films decreases up to a resistivity $\rho \sim 5 \times 10^2\,\mu\Omega\,\mathrm{cm}$. This change in the microstructure is accompanied by an increase of the superconducting gap and critical temperature measured across the macroscopic sized samples, see Fig. 2.1. Therefore, we first focus our attention on finite size effects to explain the enhanced T_c.

For the following discussion it is useful to recall the BCS theory prediction for the critical temperature of a superconducting material [112]

$$k_\mathrm{B} T_\mathrm{c} = 1.13\hbar\omega_\mathrm{D}\, e^{-1/(N(0)V)}. \tag{2.1}$$

Here, ω_D is the Debye frequency, $N(0)$ the density of states (DOS) at the Fermi energy, and V the electron-electron interaction matrix element. Following earlier results by Thomson and Blatt [123], Parmenter [124] theoretically calculated that quantization effects due to the small diameter of isolated grains could lead to an increased electron-electron interaction V and consequently enhance the critical temperature. The characteristic length scale for this effect is $L = (\lambda_\mathrm{F}^2 \xi_0)^{1/3}$, with λ_F the Fermi wavelength and ξ_0 the Pippard coherence length. For a grain size equal to $L = 6.2\,\mathrm{nm}$, a T_c twice as high as the bulk value should occur in aluminum, which is in reasonable agreement with the experimental results.

Furthermore, the calculations by Parmenter also show that $2\Delta_0/(k_\mathrm{B} T_\mathrm{c})$ increases from the BCS value of 3.5 [112] to 4 for a grain size $d \ll L$. However, for coupled grains the discrete energy levels of individual grains should broaden to bands due to interaction between grains, rendering the model potentially inapplicable for low resistivity granular aluminum films as mentioned in Ref. [115]. This contradiction was resolved by Parmenter by suggesting an effective grain volume $d_\mathrm{eff}^3 = d^3/(1-0.5t)$, with t the barrier transmissivity. As long as both d and d_eff are smaller than L, enhanced superconductivity for low resistivity films due to a size-effect could persist [124].

A second possible explanation for enhanced T_c in granular aluminum films builds on a potential surface enhancement of superconductivity. Following a suggestion by Ginzburg [125], near the surface of a superconductor-dielectric interface the electron-phonon interaction might be increased by surface phonons or a variation of the Coulomb screening. Taking on this idea and inspired by calculations of De Gennes about boundary effects in superconductors, [126] grain boundaries were regarded as internal boundaries [83, 114, 127]. Therefore, the effective electron-phonon interaction computes to an average between the bulk value and the increased value at the interface, weighted according to their respective volume. For a width of the interface region d_s smaller than half the grain size, one finds a T_c increasing $\propto 1/d$. However, in measurements of the superconducting gap one should observe a gap associated with the enhanced pairing in the interface region and one associated with the bulk volume, which is contrary to the experimentally observed sharp superconducting gap.

Finally, a number of recent theoretical works predict an increase of T_c due to the so-called shell effect, which leads to an increase of the electronic density of states $N(0)$ in small nanoclusters [128–131] [cf. Eq. (2.1)]. Similar to the states in atoms delocalized electrons in metallic clusters form a set of states which are not equidistantly spaced [129]. If the highest occupied shell in such a cluster is degenerate, this corresponds to a peak in the electronic DOS and results in an enhanced T_c.

Which of the mentioned physical processes governs the increase of the critical temperature in the nanometer sized aluminum grains of granular aluminum seems to be still under debate. However, experimental results from THz spectroscopy suggest that the enhancement of T_c in granular aluminum films is an effect of enhanced superconductivity in individual aluminum grains [84] due to the fact that T_c and Δ_0 both increase in low resistivity granular aluminum films.

In a theoretical model, which resembles the microstructure of granular aluminum films, Mayoh and García-García [132] theoretically calculated a dome-like behavior of the critical temperature as a function of the inter-grain coupling, i.e. the normal state resistivity. In their model, inspired by the experimental results of Deutscher *et al.* [83], they assume an array of clean, superconducting nanograins in an insulating matrix. The grains couple via the Josephson effect, and the grain sizes obey a Gaussian distribution with a mean $\sim 5\,\mathrm{nm}$, and variance $\sim 1\,\mathrm{nm}$. As the grAl film resistivity increases, individual grains progressively decouple, which reduces the smoothing of the increased electronic density of states due to finite size effects. The size effect leads to different T_c and Δ_0 in different grains because of their size distribution. Enhancement of the critical temperature of the array now occurs when a superconducting cluster of grains with individual critical temperatures larger than the bulk material T_c forms [132]. In this sense the proposed mechanism is similar to the percolation mechanism proposed by Abeles [133]. He suggested grains in strong electrical contact as carriers of the supercurrent, since transport via tunneling of Cooper-pairs should be suppressed due to the Coulomb blockade. As long as the typical distance between such strongly coupled links is smaller than the coherence length superconducting transport across the sample should persist.

For further insight into the exact mechanism of the enhancement of T_c local measurements of the superconducting properties of individual grains, and the extraction of the distribution of superconducting gap and critical temperature across the different grains of a granular aluminum film could complement the presently available experimental data. This additional information could potentially help to refine the theoretical model and more understanding of the enhanced critical temperature in granular superconductors might be gained.

2.3 Suppression of T_c in high resistivity granular aluminum

After discussing possible mechanisms for the increase of the critical temperature in granular aluminum we will now turn our attention to the right side of the superconducting dome. In the following section we give an overview about theoretical proposals explaining the decrease of T_c as the normal state resistivity of granular aluminum films increases above $\sim 3 \times 10^2\,\mu\Omega\,\mathrm{cm}$.

According to a number of theoretical works superconductivity in granular aluminum is suppressed by either a quenching of superconductivity in individual grains or the suppression of Josephson coupling between grains, which themselves still show superconductivity [82, 84, 132, 134–136].

Following Refs. [134–136] the discussion is based on a network of grains of size d with Josephson type coupling between grains. The free energy in the superconducting state F_s of such a system is given by [135]

$$F_s = F_n - \frac{1}{2}N(0)\Delta_0^2 d^3 - \frac{\hbar\pi\Delta_0}{4e^2R}, \tag{2.2}$$

with F_n the free energy in the normal state, $N(0)$ the density of electronic states at the Fermi energy, Δ_0 the value of the gap at zero temperature, $R = \rho/d$ the resistance of the Josephson junction [134], and ρ the normal state resistivity. Large fluctuations of the superconducting order parameter are expected for $F_n - F_s \approx k_B T_c$ [135]. Using Eq. (2.2) the condition for large fluctuations in a grain coupled via the Josephson mechanism to other grains is

$$\frac{1}{2}N(0)\Delta_0 d^3 + \frac{\hbar\pi d}{4e^2\rho} \sim 1. \tag{2.3}$$

Based on this Eq. two regimes should be distinguished. If the first term in Eq. (2.3) $\gg 1$, each grain will remain superconducting regardless of the inter-grain coupling [135]. However, in the opposite regime $(1/2)N(0)\Delta_0 d^3 \ll 1$ the Josephson coupling between grains becomes the relevant energy scale. If the coupling is strong, with the second term in Eq. (2.3) $\gg 1$, the granular superconductor retains bulk superconductivity. For weak coupling, i.e. $(\hbar\pi d)/(4e^2\rho) \lesssim 1$, the fluctuations of the superconducting order parameter will suppress superconductivity, and consequently T_c in higher resistivity granular aluminum films.

Recent experimental results suggest that superconductivity in individual grains persists but fluctuations of the phase of the superconducting order parameter become dominant [82, 84]. The measure for this collective phase-coherent state is the so-called superfluid stiffness J, which is usually much larger than the superconducting gap Δ_0. However, in superconducting granular aluminum it has been shown that J is approximately proportional to $1/\rho$ [82, 84]. Therefore, in high resistivity samples J becomes the dominant energy scale and the critical temperature is suppressed due to the loss of phase coherence at temperatures $\sim J$ [120]. This experimental observation is also in accordance with theoretical results by Mayoh and García-García [132], who theoretically obtain a dome shaped critical temperature as a function of the Josephson barrier resistance. In analogy to Josephson junction physics one could also think about the decrease of the critical temperature in terms of the RCSJ model [67]. As the individual grains decouple more and more the Josephson energy decreases. This results in a shallower trapping potential of the virtual phase particle. Consequently, macroscopic quantum tunneling and thermal activation lead to a transition to the resistive state at progressively lower temperatures.

In conclusion, experimental and theoretical investigations indicate that the dome shaped critical temperature is governed by the competition between increased critical temperature in individual grains due to finite size effects, and stronger phase fluctuations of the superconducting order parameter between grains, which eventually decrease the critical temperature of macroscopic granular aluminum samples. Similar to the question of enhanced T_c, future experiments allowing to resolve the superconducting gap in individual aluminum grains as a function of the resistivity of the granular aluminum film would complement presently available data. Combining these experimental findings might further the understanding of the superconductor to insulator transition in grAl films.

3 High impedance granular aluminum resonators

As already briefly stated in the introduction, we start by investigating the microwave frequency properties of grAl in a straightforward scheme, namely a simple $\lambda/2$ resonator. Employing this very common technique of embedding a material under investigation into a resonant circuit allows to quantify the properties from a relatively easy microwave reflection measurement.

In this chapter we describe the design and measurement results of microwave frequency resonators fabricated using grAl thin films. Section 3.1 describes the 3D sample holder geometry, the finite elements methods (FEM) simulation techniques, which were employed to design the resonant frequencies and coupling quality factors of the investigated resonators, and the different resonator geometries used. In Section 3.2 we present the measurement results, followed by a section discussing possible loss mechanisms limiting the internal quality factor of the resonators. Finally, Section 3.4 discusses the dynamics of quasiparticles in grAl.

3.1 Design

The simulation tools utilized for the design of microwave resonators fabricated from grAl should naturally be adapted to the electro-magnetic (em) environment surrounding the sample. Since in this thesis we use a 3D geometry sample holder [102], which provides a comparably clean em environment, we start this section with a brief description of the sample holder geometry. Following this discussion we present the approach used to simulate materials with non-negligible kinetic inductance and finally discuss the realized designs.

3.1.1 Sample holder

The tremendous increase in coherence times of superconducting circuits over the last years has been - in large part - attributed to the optimization of fabrication techniques and optimized geometries [2]. Improved cleaning and surface treatment processes have led to dielectric loss tangents of interfaces (cf. Sec. 1.4.2), which today are in the range of $\tan(\delta) \sim 10^{-3}$ [91–93]. To mitigate the effect of these dissipative circuit regions, a possible strategy is the increase of the resonant mode volume. This is accompanied by a reduction of the electric field strength, which in turn leads to a decreased participation ratio of the interface regions and hence increases the internal quality factors of such circuits. Popular examples of this technique are the 3D transmon [32, 137] or 3D cavity resonators for quantum information with internal quality factors approaching 10^8 [138], while resonant cavities with similar geometries to ones used for applications in accelerators even achieve internal quality factors $Q_\mathrm{i} \sim 10^{10}$ [139].

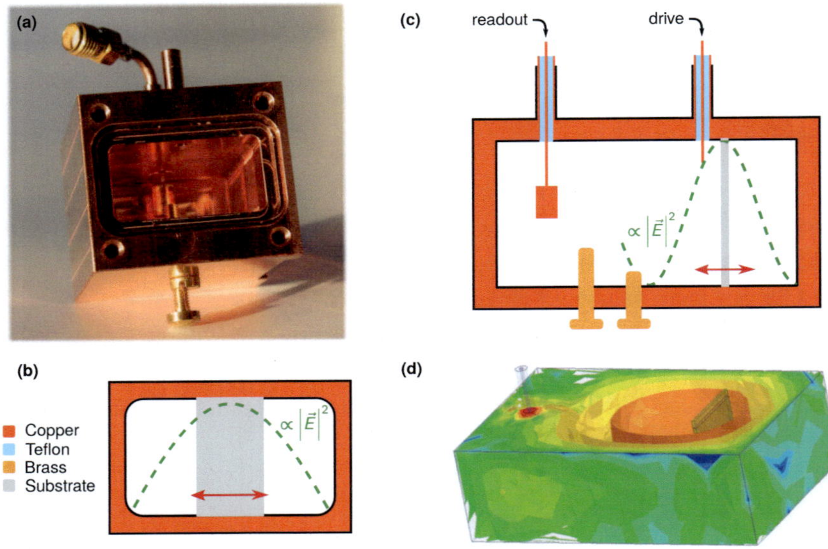

Figure 3.1: Image and schematic drawing of a 3D rectangular waveguide sample holder. **(a)** Photograph of a waveguide sample holder with the capping lid removed. At the far end inside the waveguide the pin with barrel is visible, which provides impedance matching between the waveguide and 50 Ω coaxial lines. The two brass screws used for tuning the matching are visible on the bottom of the waveguide. A second input port is visible on the top of the waveguide towards the front. Here, a second coax cable can be inserted without a matching barrel to drive circuits at frequencies below the waveguide cutoff. **(b, c)** Schematic front (b) and side (c) views of the sample holder with indication of the electric field distribution (green dashed line). Red arrows indicate possible directions to adjust the dipole coupling to the waveguide's TE_{10} mode (see main text). **(d)** FEM simulation results schematically illustrating the electric field amplitude distribution of the TE_{10} mode (red, orange: large electric field, blue, green: small electric field).

In the spirit of the previous paragraph, we follow the approach of A. Kou *et al.* [102] who use a 3D rectangular wave guide, which is matched to the 50 Ω coax lines of the cryogenic measurement setup. Therefore, it acts as an extension of the transmission line with a shorted end. Figure 3.1 shows a picture of a copper wave guide sample holder, a schematic drawing of the geometry, and an illustration of the magnitude of the electric field. The wave guide has a fundamental cutoff frequency $\sim 6\,\mathrm{GHz}$, while the second waveguide mode propagates frequencies $\gtrsim 8\,\mathrm{GHz}$. This leaves a well behaved, single-mode frequency band between 6 GHz to 8 GHz, in line with commercially available microwave components. By optimizing, via FEM simulations, the position and dimensions of the coupling ports and impedance matching screws we achieve a return loss $\lesssim -10\,\mathrm{dB}$ in the band.

After mounting the sample, the waveguide is closed with a copper lid, which is sealed with an indium wire ($\sim 0.1\,\mathrm{mm}$ in diameter). This lid shorts the waveguide at $\sim \lambda/4$ of $7\,\mathrm{GHz} \approx 1\,\mathrm{cm}$ away from the sample position, which results in an electric field anti-node, and allows strong coupling of an electric dipole to the waveguide's propagating field. For measurements where a weak coupling to the transmission line is of interest, the waveguide in principle offers three geometric and one frequency based method to achieve almost arbitrarily high coupling quality factors (see Fig. 3.1) Moving the sample with respect to the shorted end of the waveguide, or moving the sample laterally, results in a smaller electric field interacting with the sample's dipole moment, which results in a smaller coupling. Furthermore, rotating the electric dipole with respect to the electric field will decrease the interaction and will result in a complete decoupling for a perpendicular orientation between dipole and electric field. Lastly, designing the samples characteristic frequency below the waveguide cutoff frequency is another approach to achieve coupling quality factors in the range of 10^6 (cf. Ref. [95]), due to the evanescent propagation of microwaves with frequencies below the waveguide cutoff.

3.1.2 FEM simulation of designs employing kinetic inductance

Owing to our use of a rectangular waveguide as the sample holder we perform full 3D FEM simulations in order to design resonators from grAl at a desired resonant frequency and specific coupling quality factor. To achieve this, we use a full 3D model of the waveguide in ANSYS HFSS™, which also includes the impedance matching section, see Fig. 3.2. In the following we will briefly outline the simulation approach used to simulate structures with considerable kinetic inductance.

For a simulation of the sample, we either import a GDSII layout file and move the structures onto the plane of the virtual substrate or model the parametrized geometry directly in HFSS. We represent any metal thin film as zero-thickness sheets, since for our simulation purposes the thickness does not play a role, and would only significantly increase the computational cost of the simulations due to the differences in dimensions between the film thickness (20 nm) and the waveguide ($\sim 10\,\mathrm{mm}$).

To obtain the resonant frequencies of a certain design implemented into the HFSS model, we employ the program's *eigenmode* solver. The key features of this solver are its capability to return not only the frequencies of eigenmodes but also their corresponding quality factors. In contrast to a driven modal solver only one simulation run is necessary. It does not however, give access to the frequency dependence of the impedance of the modeled structure.

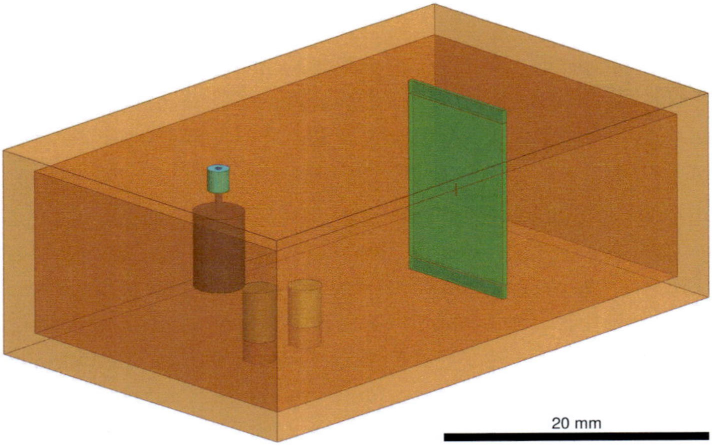

Figure 3.2: Model for the full 3D simulation of the sample holder in ANSYS HFSS. The entire waveguide including the impedance matching section consisting of the pin with matching barrel (far left), and the adjusting screws (cylinders on the bottom) is simulated. Light blue indicates the dielectric of the 50 Ω coax line and is terminated with a matched load. This approach allows to efficiently simulate the coupling between the waveguide sample holder and the designed circuit, which is realized as sheets with corresponding boundary conditions on the sapphire substrate (circuit not visible, sapphire substrate - green rectangle).

In order to include a kinetic inductance into the simulation of a wire, we use the *surface impedance boundary* option of sheets, which allows to specify a value per square, i.e. $Z_s = \omega_0 L_{kin}$ with ω_0 the expected resonant frequency and L_{kin} the designed value of the kinetic inductance. The result of the eigenmode solver now serves as a first approximation for a subsequent iterative process, where we adjust the frequency ω_0 used to specify the surface impedance until it matches the result of the eigenmode solver. Identifying the design's resonance among the solutions of the solver is also facilitated by this iterative process, since it allows to pinpoint the modes depending on the kinetic inductance. Furthermore, the display of surface currents or electric field simulations is a valuable tool in identifying the actual resonant frequency of the design among all eigenmode solver results.

By simulating our entire rectangular waveguide sample holder including a part of the coaxial line, which we terminate in a 50 Ω load in the simulation, we also gain access to the coupling quality factor of the simulated design via the Q-factor result of the eigenmode solver. Setting all materials besides the substrate as lossless perfect conductors, all dissipation occurs in the virtual load of the coaxial line. Under the assumption that the internal quality factor in the simulation approaches infinity, since the simulated losses in the substrate are negligible compared to the usually targeted coupling quality factors $\lesssim 10^5$, we observe $1/Q_{\text{sim}} = 1/Q_{\text{i}} + 1/Q_{\text{c}} \approx 1/Q_{\text{c}}$. Therefore, the eigenmode solver gives direct access to the coupling quality factor, provided the internal losses integrated into the simulation model are negligible and the coax port is terminated in a matched $50\,\Omega$ load.

In fully planar designs it mostly suffices to simulate only the substrate with metal patterns and the corresponding dielectrics above and below. Such a 2.5D simulation is computationally often more efficient and time saving than a full 3D simulation. Therefore, preliminary simulations during an exploratory stage of the design process could be carried out as 2.5D simulations. One possible simulation software for such tasks is SONNET™. It also allows to straightforwardly implement a material with the desired kinetic inductance. However, the program does not supply an eigenmode solver and only the S-parameters of software defined ports are simulated. Using a fitting algorithm, the parameters of interest (f_0 and Q_{c}) are accessible from the full frequency spectrum. In case 2.5D simulations were employed for samples to be measured in the rectangular waveguide sample holder geometry, we crosscheck the preliminary design in a complete 3D simulation and refine there if necessary.

3.1.3 Rectangular strip resonators

The first geometry we design for the evaluation of grAl regarding its suitability to realize superinductors is a planar resonator. For reasons outlined in Sec. 3.1.1 we employ a 3D rectangular waveguide as our sample holder, which reduces the participation ratio of lossy interface regions (cf. Sec. 1.4.2 and Refs. [90–93]). In order to allow direct coupling of the resonator to the waveguide mode, we use a simple $\lambda/2$ design, which we align parallel to the electric field of the TE_{10} mode of the sample holder. The chip does not have a backside metalization, since this would reduce the resonators mode volume and introduce unwanted scattering of the $\vec{E}$-field in the waveguide. Consequently, the geometry chosen for this first set of samples is a simple rectangle fabricated from a 20 nm thick grAl film. To avoid confusion with standard microwave taxonomy, we call this geometry *rectangular strip* or antenna geometry. The overall approach is similar to a standard microstrip setup, albeit the missing ground plane on the backside of the substrate.

Based on previously presented results about the microwave properties of grAl resonators by Rotzinger *et al.* [140], we target a sheet resistance $R_\square \sim 1\,\text{k}\Omega$, corresponding to an expected kinetic inductance $L_{\text{kin}} \approx 0.8\,\text{nH}/\square$. We place three resonators on a $10\,\text{mm} \times 15\,\text{mm}$ c-plane sapphire chip, mainly to cover a range of about two orders of magnitude in coupling quality factors, and to be able to cross check results between the different resonators. To allow for reliable extraction of the internal quality factors of the resonators, we aim for coupling quality factors Q_c on the order of the internal quality factor Q_i. Previous experiments in a coplanar waveguide geometry (CPW) have shown internal quality factors of grAl resonators $Q_i \sim 10^4 - 10^5$. Hence, for our experiment we design couplings $Q_c \sim 10^3$, $\sim 10^4$, and $\sim 10^5$.

Table 3.1: To maximize fabrication success in the presence of parameter fluctuation, we design resonators with half and twice as large kinetic inductance. We obtain three sets of resonators, which are implemented on the same chip and each fabricated twice within each run. In order to be able to fit a wide range of internal quality factors reliably, we cover two orders of magnitude in Q_c to have resonators with $Q_c \sim Q_i$ for internal quality factors up to 10^6.

Q_c	L_{kin} (nH)	Dimension (μm^2)
10^3	0.4	10×1000
10^4	0.4	2.5×600
10^5	0.4	1.35×400
10^3	0.8	20×1000
10^4	0.8	5×600
10^5	0.8	2.7×400
10^3	1.6	40×1000
10^4	1.6	10×600
10^5	1.6	5.4×400

Using the FEM simulation method described in Sec. 3.1.2, we aim for resonant frequencies of 6.0 GHz to 7.5 GHz. Table 3.1 shows the designed resonator dimensions for the targeted kinetic inductance of 0.8 nH/□, and values 50% larger and smaller, to cover a potential spread in obtained sheet resistances observed in earlier fabrication runs (cf. Sec. A.1). For device fabrication, we use 2" sapphire wafers, which fit six chips with dimensions 10 mm × 15 mm. Based on this, we place each set consisting of three resonators designed for the same kinetic inductance on one chip and fabricate every chip twice (cf. Table 3.1). See Figure 1.11 for an optical image of a chip with a set of rectangular strip resonators designed for a kinetic inductance $L_{kin} = 0.8$ nH/□.

3.1.4 Microwave kinetic inductance detectors

Thanks to its large kinetic inductance, aside from applications for superconducting quantum circuits, grAl is also of interest for microwave kinetic inductance detectors (MKID) [99]. In these sensors an element with some kinetic inductance is shunted with a capacitor to form an *LC* microwave resonator. However, the absorption of photons with frequency $\hbar\omega \geq 2\Delta_0$ leads to the breaking of Cooper pairs, which increases the kinetic inductance and hence leads to a shift in frequency for the resonant circuit. Because only the kinetic inductance is susceptible to the change in Cooper pair density, a larger signal is achieved for circuits with higher kinetic inductance fraction $\alpha = L_{\text{kin}}/(L_{\text{geo}} + L_{\text{kin}})$. The performance of MKIDs fabricated from grAl thin films was investigated recently and is described by Valenti *et al.*, see Ref. [141]. Nevertheless, measurements of MKID-type resonators in a well shielded setup provide some useful insights into the properties of grAl regarding its use in superconducting quantum circuits.

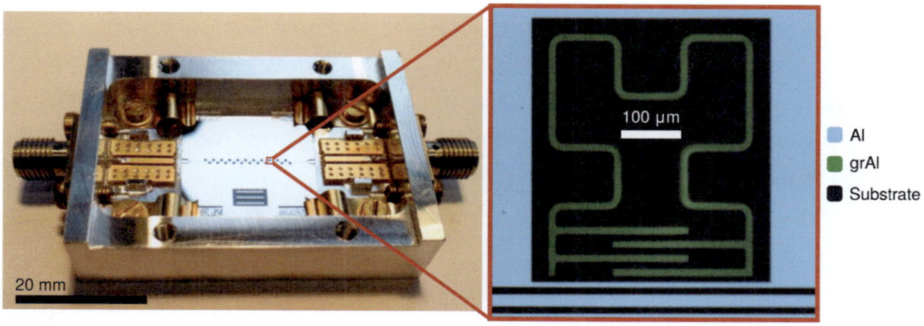

Figure 3.3: Image of MKID sample holder and false colored zoom-in showing one resonator in detail. The central transmission line on the chip is connected to the cryogenic measurement setup via wirebonds and printed circuit boards (PCBs) on both sides of the sample box. Four springs hold the chip in place and press it down to provide thermal contact. The individual resonators (green) consist of an interdigitated capacitor and an inductor, which is formed as a Hilbert fractal [142]. All resonators are fabricated from a grAl thin film using an *e*-beam lift-off process (cf. Appendix 3). The ground plane and transmission line are patterned in a second optical lithography lift-off step from a pure aluminum thin film (blue).

MKIDs measured in this thesis have been designed by the group of A. Monfardini at Institut Néel in Grenoble, who successfully employ such geometries for radioastronomy applications [100]. The fabrication of all MKID type samples was carried out at KIT with a two step lithography process. In a first step we pattern the resonators from grAl thin films. In a second step we add the pure aluminum ground plane and feedline of the chip. Figure 3.3 shows the sample holder used to measure the MKID chips and a zoom-in, which shows a typical MKID resonator. To adjust the resonant frequency, we vary the width of the inductor between different chips depending on the design value for the kinetic inductance. For resonators on the same chip the inductor width is constant, while we vary the capacitance of the interdigitated finger capacitor. Each chip consists of 22 resonators capacitively coupled in a hanger geometry to a common feedline, which is wirebonded to a PCB. SMA plugs are soldered onto the PCB and allow to connect the sample to the cryogenic microwave measurement setup.

3.1.5 Coplanar waveguide resonators

Coplanar waveguides and resonators are commonly used in superconducting quantum circuits for readout purposes and to implement coupling between distant qubits. Owing to their smaller mode volume than 3D approaches, the participation ratios of interfaces are higher and the resulting dielectric loss has proven to be a limiting loss mechanism for coherence [90–93]. As a cross check, we evaluate the internal quality factor in the single photon regime of CPW grAl resonators as a function of their interface participation ratio. These resonators have been fabricated by sputter deposition of 20 nm thick grAl films with sheet resistivity $R_{\square} \sim 1\,\mathrm{k}\Omega$ and a subsequent chlorine based dry etching process (see Ref. [140] for details). The resonators are $\lambda/4$ resonators and capacitively coupled to a transmission line, which is probed in transmission. Table 3.2 provides an overview over the gap and center strip dimensions of the CPW as well as the resonant frequency.

Table 3.2: Gap and centerstrip width, as well as resonant frequency of three CPW resonators fabricated by sputter deposition of a 1 kΩ grAl film and subsequent dry etching. The resonators are designed with a length of $\lambda/4$, see Ref. [140] for further details.

Label	Gap (μm)	Center strip (μm)	f_r (GHz)	Q_i (single photon regime)
CPW1	2	1.2	5.3	4×10^4
CPW2	4.8	8.0	7.3	2.4×10^5
CPW3	9.6	16.0	8.0	2.6×10^5

3.2 Measurement results

In this section we show measurement results of the different grAl thin film resonators investigated over the course of this PhD thesis. These results serve as a basis for a subsequent interpretation and discussion of the results, which is carried out in Sec. 3.3. The presentation of the measured data is organized by their resonator geometry, with Sec. 3.2.1 giving an overview of rectangular strip resonators measured in the 3D waveguide sample holder, while Sec. 3.2.2 describes measurements of the MKID resonators in a fully planar design.

3.2.1 Rectangular strip resonators

All rectangular strip resonators are measured in a 3D waveguide sample holder geometry (cf. Sec. 3.1.1) and designed according to the procedure outlined in Sec. 3.1.2. The well controllable electro-magnetic environment, together with an increased mode volume, should reduce in principle the susceptibility to dielectric and radiative losses. Since a goal of this thesis is to characterize granular aluminum for its suitability as a material of choice for superconducting quantum circuits, we aim at a sample environment which allows high internal quality factors in order to achieve high sensitivity to potential added or stronger loss mechanisms intrinsic to grAl resonators.

In the first two subsections we present results of two resonators fabricated from a 20 nm thick 1.4 kΩ film, using a chlorine based dry etching process of the sputter deposited grAl (see Ref. [140] for details of the sputtering process). The following sections detail measurement results of three resonators fabricated in *e*-beam lift-off lithography on c-plane sapphire substrates using grAl films deposited in a Plassys™ *e*-beam evaporation machine (cf. Appendix 3 and Appendix A.1).

Resonator A1: $R_\square = 1.4\,\mathrm{k\Omega/\square}$, $t = 20\,\mathrm{nm}$

Sample A1 is one of four resonators fabricated on a 15 mm × 15 mm c-plane sapphire chip, on which a $R_\square = 1.4\,\mathrm{k\Omega}$ grAl film with thickness $t = 20\,\mathrm{nm}$ was deposited by a sputtering process [140]. Due to the fact that a single optical mask was designed for multiple film resistivities, and a chip size of 10 mm × 15 mm, instead of the targeted three resonators, the used larger chip holds five.

Furthermore, during the measurement run on this chip only two resonators could be identified and were thoroughly characterized. Therefore, the unambiguous assignment of resonator geometry to measured data is not possible for this sample. Hence, we cannot extract the kinetic inductance of the resonators by comparison with FEM simulations. However, the resonant frequency of sample A1 $f_0^{A1} = 7.2317\,\mathrm{GHz}$ is in the targeted range, indicating that the kinetic inductances should be within $\sim 50\%$ of the expected value of $L_{\mathrm{kin}} \sim 1.1\,\mathrm{nH/\square}$.

Figure 3.4 shows the measured amplitude data as a function of the readout power applied with the VNA. The quoted power is an estimate based on the total attenuation in the cryogenic measurement setup (see Appendix C.1) and the power emitted by the VNA. As can be seen by the center panel of the plot, the internal quality factor as a function of the average number of photons circulating in the resonator decreases from a value on the order of 10^5 at high photon numbers, to $\sim 6 \times 10^4$ in the single photon regime. The total and coupling quality factors as a function of the average number of photons are given in the bottom left panel of Fig. 3.4. In sample A1 the total quality factor is limited by the coupling quality factor with its mean value $\overline{Q_\mathrm{c}} = 11 \times 10^3$ (orange points) and a scattering between points due to the decreasing signal to noise ratio (SNR) at smaller average photon numbers.

Finally, the bottom right panel of Fig. 3.4 shows the resonant frequency as a function of $\bar{n}$ in log-lin scale. The orange line is a linear fit to the resonant frequency from which we extract a self-Kerr coefficient $K_{11} = 6\,\text{Hz}$. Since we cannot perform direct photon number calibration but rely on the estimated total attenuation, we consider $\bar{n}$ to be accurate within one order of magnitude. Therefore, the accuracy of the self-Kerr coefficient is of the same order.

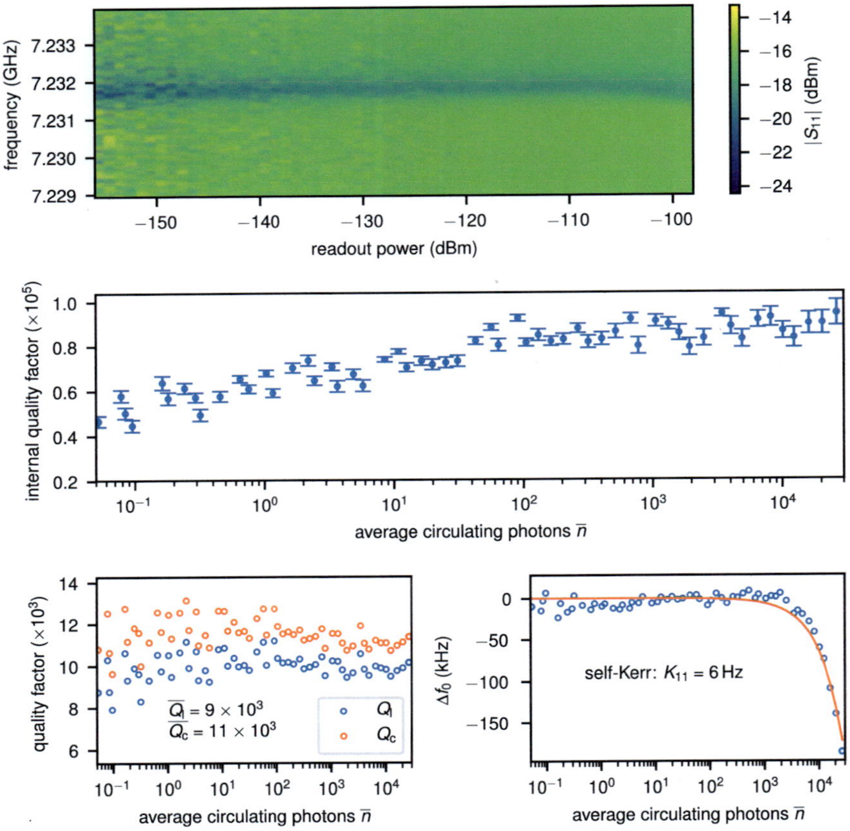

Figure 3.4: Measurement results of sample A1, fabricated from a grAl film with $R_\square = 1.4\,\text{k}\Omega$ and a thickness of 20 nm. The resonant frequency of the sample is 7.2317 GHz and the top panel shows as a color plot the raw amplitude signal. In the center panel the internal quality factor Q_i is plotted as a function of the average number of photons circulating in the resonator $\bar{n}$. In the single photon regime, we extract $Q_\text{i} \sim 6 \times 10^4$. The total quality factor is limited by the coupling quality factor, whose mean value $\overline{Q_\text{c}} = 11 \times 10^3$ (see bottom left panel). From a linear fit to the resonant frequency as a function of $\bar{n}$ we extract a self-Kerr coefficient $K_{11} = 6\,\text{Hz}$.

Figure 3.5 shows the raw data (blue points) and fits of the resonator reflection measurement at a sample holder input port power $P = -142\,\text{dBm}$, corresponding to an average number of circulating photons in the resonator $\bar{n} \approx 1$. As can be seen, the fit matches the measured data well, and the internal quality factor is about a factor of 6 higher than the coupling quality factor, which allows a reliable extraction of Q_i (cf. error bars in Fig. 3.4).

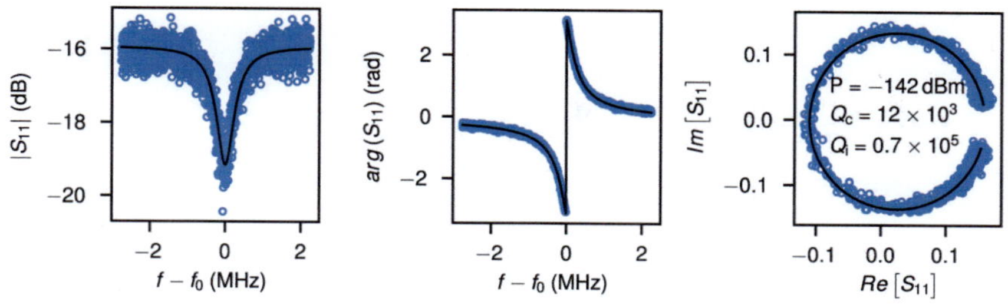

Figure 3.5: Raw data (blue points) and circle fit (black line) of sample A1 in the single photon regime. The fit is in good agreement with the measured data and we extract $Q_i = 7 \times 10^4$ at a coupling quality factor $Q_c = 12 \times 10^3$ at a readout power applied to the waveguide sample holder input port of $P = -142$ dBm, which corresponds to $\bar{n} \approx 1$.

Resonator B1: $R_\square = 1.4\,\mathrm{k\Omega/\square}$, $t = 20\,\mathrm{nm}$

Sample B1 is the second identified and in detail measured resonator fabricated by optical lithography and a chlorine based dry etching process (cf. Subsection 3.2.1). Due to the reasons already outlined in the previous section, an unambiguous identification of the resonator geometry is not possible for Sample B1 either.

Figure 3.7 shows again the measured amplitude data as a function of the readout power incident on the matched sample holder port. At low readout powers, we measure a resonant frequency $f_0^{B1} = 7.6139$ GHz. In the single photon regime we extract an internal quality factor $Q_i \sim 2 \times 10^4$ (see center panel), which is comparable to the coupling quality factor $Q_c = 22 \times 10^3$ (see bottom left panel). From the shift of the resonant frequency with increasing number of average circulating photons $\bar{n}$ we extract a self-Kerr coefficient 15 Hz.

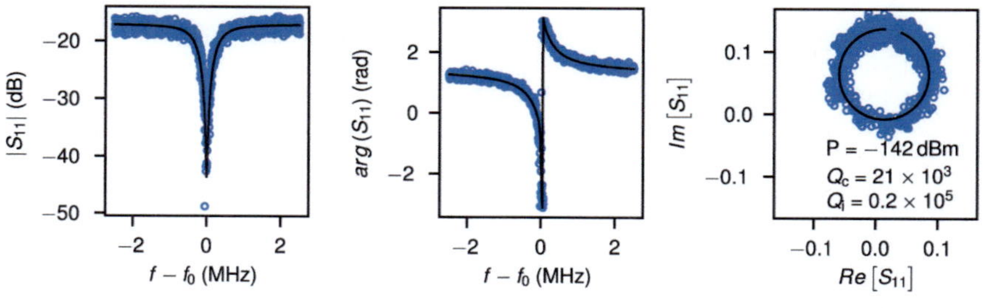

Figure 3.6: Raw data (blue points) and circle fit (black line) of sample B1 in the single photon regime. The fit is in good agreement with the measured data and we extract $Q_i = 2 \times 10^4$ at a coupling quality factor $Q_c = 21 \times 10^3$ at a readout power applied to the waveguide sample holder input port of $P = -142$ dBm, which corresponds to $\bar{n} \approx 1$.

Figure 3.6 shows the raw data (blue points) for a readout power of −142 dBm, which corresponds to approximately one photon on average circulating in the resonator. As expected for almost critical coupling $Q_\mathrm{i}/Q_\mathrm{c} \approx 1$, we observe a deep dip in the amplitude data, and a circle of the complex reflection coefficient S_{11} which passes the origin of the real-imaginary plane.

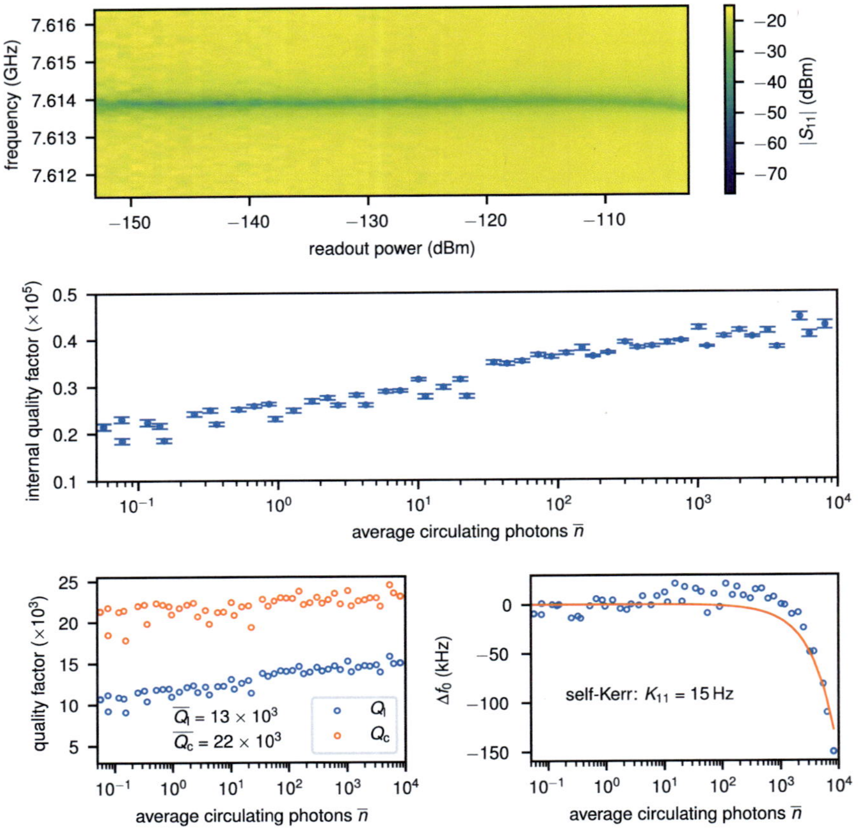

Figure 3.7: Measurement results of sample B1, fabricated from a grAl film with $R_\square = 1.4\,\mathrm{k\Omega}$ and a thickness of 20 nm. The resonant frequency of the sample is 7.6139 GHz and the top panel shows as a color plot of the raw amplitude signal. In the center panel the internal quality factor Q_i is plotted as a function of the average number of photons circulating in the resonator $\overline{n}$. In the single photon regime, we extract $Q_\mathrm{i} \sim 2 \times 10^4$. The mean value of the coupling quality factor $\overline{Q_\mathrm{c}} = 22 \times 10^3$ (see bottom left panel). From a linear fit to the resonant frequency as a function of $\overline{n}$ we extract a self-Kerr coefficient $K_{11} = 15\,\mathrm{Hz}$.

Resonator A2: $R_\square = 2.0\,\mathrm{k\Omega/\square}$, $t = 20\,\mathrm{nm}$

Sample A2 is the weakest coupled resonator of the set of three resonators fabricated on the same $10\,\mathrm{mm} \times 15\,\mathrm{mm}$ c-plane sapphire chip. These resonators are patterned using the *e*-beam lift-off lithography process described in Appendix A.3). Due to the clear differences in coupling quality factors we can assign sample A2 to a resonator with dimensions 400 µm × 5.4 µm. Assuming the fabricated resonator has exactly these dimensions, we extract a kinetic inductance of $L_{\mathrm{kin}} = 2\,\mathrm{nH}/\square$ by comparing the measured resonant frequency $f_0^{A2} = 6.9945\,\mathrm{GHz}$ with FEM simulations, where we sweep the kinetic inductance to match the observed frequency.

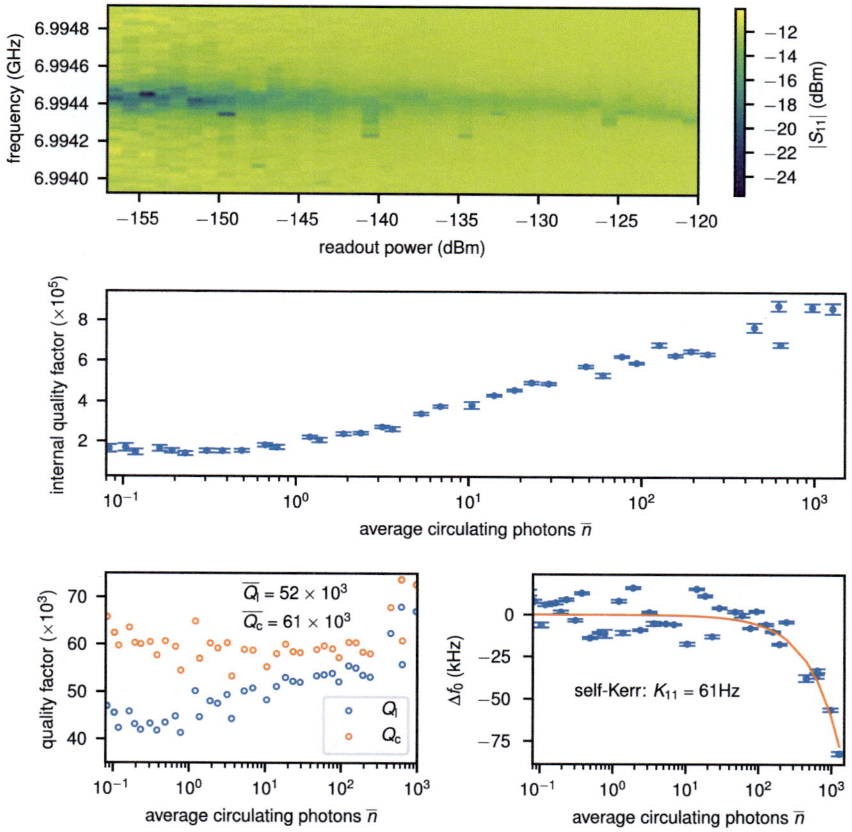

Figure 3.8: Measurement results of sample A2, fabricated from a grAl film with $R_{\square} = 2\,\mathrm{k\Omega}$ and a thickness of 20 nm. The resonant frequency of the sample is 6.9945 GHz and the top panel shows as a color plot the raw amplitude signal. In the center panel the internal quality factor Q_{i} is plotted as a function of the average number of photons circulating in the resonator $\overline{n}$. In the single photon regime, we extract $Q_{\mathrm{i}} \sim 2 \times 10^5$, and a mean value of the coupling quality factor $\overline{Q_{\mathrm{c}}} = 61 \times 10^3$ (see bottom left panel). From a linear fit to the resonant frequency as a function of $\overline{n}$ we extract a self-Kerr coefficient $K_{11} = 61\,\mathrm{Hz}$.

As discussed in the introduction we attribute the discrepancy between the extracted kinetic inductance and the one expected from Mattis-Bardeen formula to differences between the designed and realized resonator geometry, film inhomogeneity across the 2” wafer, or a deviation from the standard Mattis-Bardeen formula, since at such high resistivities the ratio $\Delta_0/(k_B Tc)$ differs from the standard BCS value of 1.764 and is ~ 2.1 [82].

Figure 3.8 shows the raw amplitude data as a function of the VNA readout power reaching the matched port of the 3D waveguide sample holder. The low amplitude points away from the resonance dip result from frequency fluctuations due to quasiparticle producing events, which increase the kinetic inductance temporarily (cf. Sec. 1.4.3 and 3.4). Since these impacts distort the amplitude and phase spectrum of the resonator, measurements where a quasiparticle burst occurred are not fitted. Despite the impacts we extract internal quality factors between 8×10^5 and 1×10^5 in the high and single photon regime for undisturbed measurements, respectively. The average coupling quality factor $\overline{Q_c} = 6 \times 10^4$ (see bottom left panel in Fig. 3.8) is on the order of the single photon Q_i and limits the total quality factor.

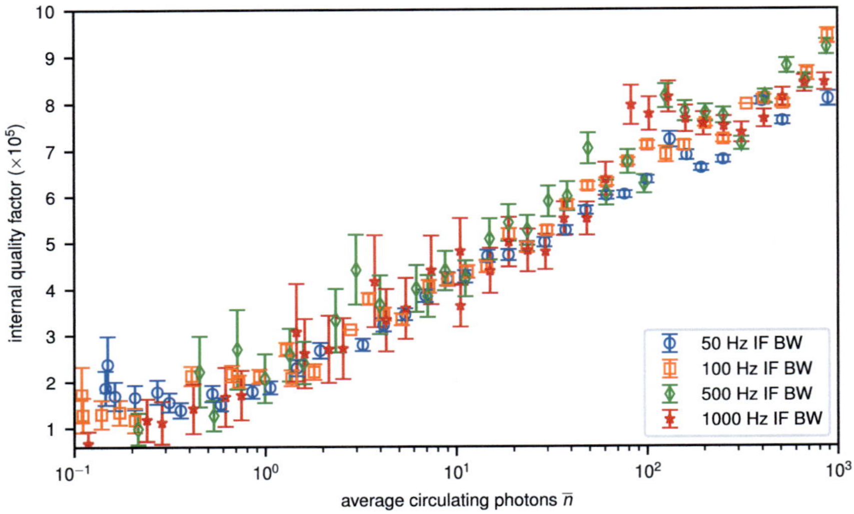

Figure 3.9: Internal quality factor as a function of the average number of circulating photons at different averaging times. Internal quality factors fitted from data with averaging times up to 100 times shorter (1000 Hz IF BW) than the results presented in Fig. 3.8 show comparable values. Therefore, we conclude that the extracted values are not increased due to an averaging out of the dip in amplitude signal by a jittering resonant frequency.

In principle, a jittering resonator frequency on time scales faster than a measurement would lead to the averaging of multiple resonator signatures at slightly different frequencies. This would result in a shallower dip for the amplitude signal and could potentially result in higher fit values for the internal quality factor. As a control experiment, we perform measurements with swept readout power at increasing measurement speed, i.e. shorter averaging, see Fig. 3.9. In practice, we achieve this by increasing the intermediate frequency bandwidth (IF BW) of the VNA. The initial measurements presented in Fig. 3.8 are performed using an IF BW of 10 Hz. Measurements with IF BW of 100 Hz and 1000 Hz, which correspond to

10 respectively 100 times shorter averaging time show similar internal quality factors as observed for longer measurements, see Fig. 3.9. Therefore, we conclude that the extracted internal quality factors are not a result of an effective averaging out of a deeper amplitude dip due to a fluctuating resonant frequency.

A linear fit of the resonant frequency plotted as a function of $\overline{n}$ yields a self-Kerr coefficient $K_{11} = 61\,\mathrm{Hz}$. We consider this result accurate within a factor of ten due to the uncertainty in the estimation of the photon number and because only a small frequency shift could be observed, before the resonator shows bifurcation behavior. Figure 3.10 shows the good agreement between the fitted model (black line) and the measured data (blue points) for a power corresponding to $\overline{n} \approx 1$.

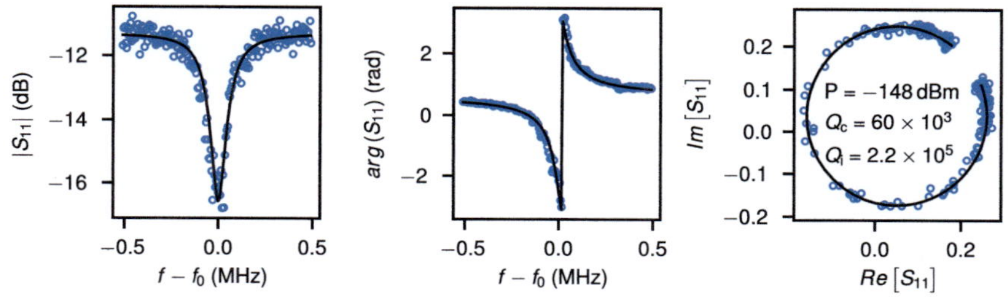

Figure 3.10: Raw data (blue points) and circle fit (black line) of sample A2 in the single photon regime. The fit is in good agreement with the measured data and we extract $Q_i = 2.2 \times 10^5$ at a coupling quality factor $Q_c = 60 \times 10^3$ at a readout power applied to the waveguide sample holder input port of $P = -148\,\mathrm{dBm}$, which corresponds to $\overline{n} \approx 1$.

Resonator B2: $R_\square = 2.0\,\mathrm{k}\Omega/\square$, $t = 20\,\mathrm{nm}$

Sample B2 is the strongest coupled resonator of the set fabricated from a $2\,\mathrm{k}\Omega$ grAl film with thickness 20 nm. Again comparing the measured resonant frequency $f_0^{B2} = 6.0324\,\mathrm{GHz}$ with FEM simulations of its designed dimensions $1000\,\mu\mathrm{m} \times 40\,\mu\mathrm{m}$, we confirm the kinetic inductance of $L_{\mathrm{kin}} = 2\,\mathrm{nH}$ extracted from sample A2. Due to its larger line width and shallower depth no impacts are visible in the power dependent measurement of the resonator's amplitude signal shown in Fig. 3.11. We observe internal quality factors on the order of 10^5 in the single photon regime, which only increase up to $\sim 2 \times 10^5$ at high powers. In sample B2 the total quality factor is entirely limited by the coupling quality factor $Q_c \approx 4 \times 10^3$, as can be seen in the bottom left panel of Fig. 3.11. We observe a small frequency shift as a function of increasing power on the order of 20 kHz, which is significantly smaller than the resonator line width (cf. Fig. 3.12). In combination with the uncertainty of the average number of circulating photons, $K_{11} = 25\,\mathrm{Hz}$ should be treated as an order of magnitude estimate for the self-Kerr coefficient.

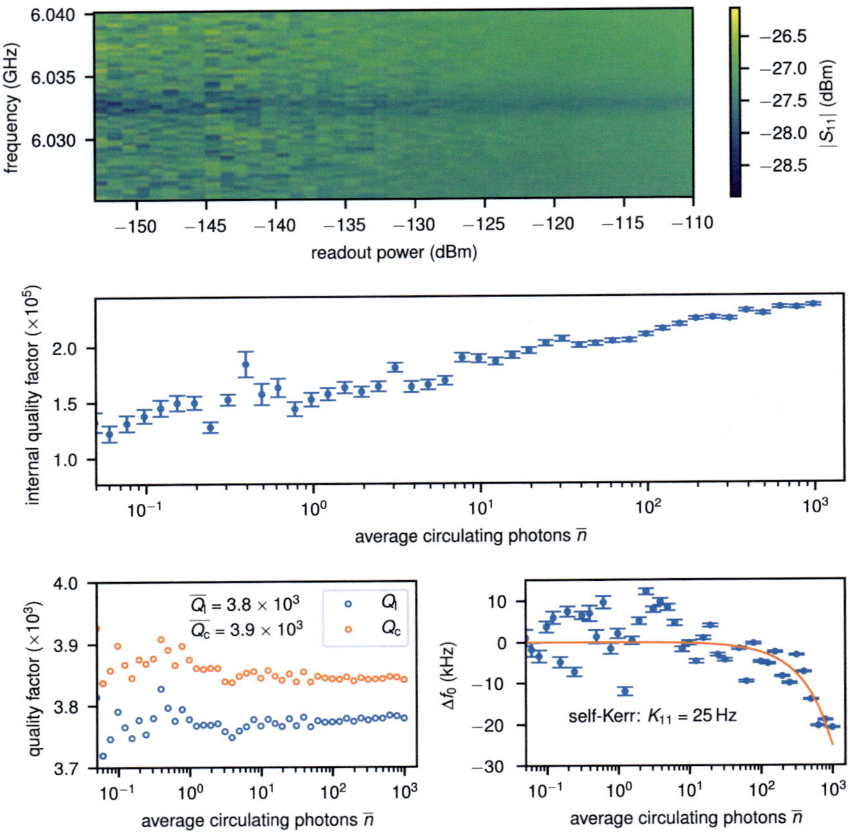

Figure 3.11: Measurement results of sample B2, fabricated from a grAl film with $R_\square = 2\,\mathrm{k}\Omega$ and a thickness of 20 nm. The resonant frequency of the sample is 6.0324 GHz and the top panel shows as a color plot the raw amplitude signal. In the center panel the internal quality factor Q_i is plotted as a function of the average number of photons circulating in the resonator $\overline{n}$. In the single photon regime, we extract $Q_\mathrm{i} \sim 1.5 \times 10^5$. The total quality factor is limited by the coupling quality factor, whose mean value $\overline{Q_\mathrm{c}} = 4 \times 10^3$ (see bottom left panel). From a linear fit to the resonant frequency as a function of $\overline{n}$ we extract a self-Kerr coefficient $K_{11} = 25\,\mathrm{Hz}$.

Figure 3.12 shows the fit (black line) to the raw data (blue points) in the single photon regime. As can be seen, a significant scatter of the amplitude data is visible. The phase response however exhibits a clean signal and is well matched by the fitted model. From the circle fit procedure we extract an internal quality factor $Q_\mathrm{i} = 1.6 \times 10^5$ at a readout power of $-139\,\mathrm{dBm}$, which corresponds to $\overline{n} \approx 1$.

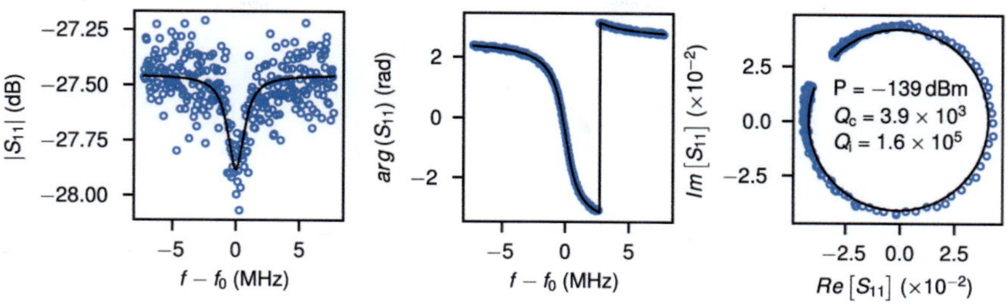

Figure 3.12: Raw data (blue points) and circle fit (black line) of sample A1 in the single photon regime. The fit is in reasonable agreement with the measured amplitude data and in good agreement with the measured phase data. We extract $Q_\mathrm{i} = 1.6 \times 10^5$ at a coupling quality factor $Q_\mathrm{c} = 4 \times 10^3$ at a readout power applied to the waveguide sample holder input port $P = -139\,\mathrm{dBm}$, which corresponds to $\bar{n} \approx 1$.

Resonator C2: $R_\square = 2.0\,\mathrm{k}\Omega/\square$, $t = 20\,\mathrm{nm}$

Sample C2 exhibits some interesting behavior at its resonant frequency $f_0^{C2} = 6.3308\,\mathrm{GHz}$, resulting from a $600\,\mu\mathrm{m} \times 10\,\mu\mathrm{m}$ rectangular strip resonator fabricated from the identical grAl film also used to pattern resonators A2 and B2. As can be faintly seen in the plot of the raw amplitude data in Figure 3.13 the signal changes from a peak at high readout powers (center panels) to the expected dip at low readout powers (bottom panels).

The powers corresponding to the line cuts shown in the 2nd and 3rd row of Fig. 3.13 are indicated by dashed lines in the amplitude plot with their respective colors. Using the standard circle fit routine also employed for samples A2 and B2 at a power corresponding to $\bar{n} \approx 0.1$ (orange), we extract a coupling quality factor $Q_\mathrm{c} = 8 \times 10^3$ and an internal quality factor $Q_\mathrm{i} = 2.6 \times 10^5$. Assuming a constant coupling quality factor, which limits the loaded quality factor $Q_\mathrm{l} \approx Q_\mathrm{c}$, we estimate that the resonator shows a clearly visible peak in its amplitude signal at $\bar{n} \approx 100$, corresponding to a readout power of $-123\,\mathrm{dBm}$.

We attribute the observed behavior of the amplitude signal to impedance mismatches in our cryogenic measurement setup or the interaction of the resonator with a spurious mode close in frequency to the resonant frequency of sample C2.

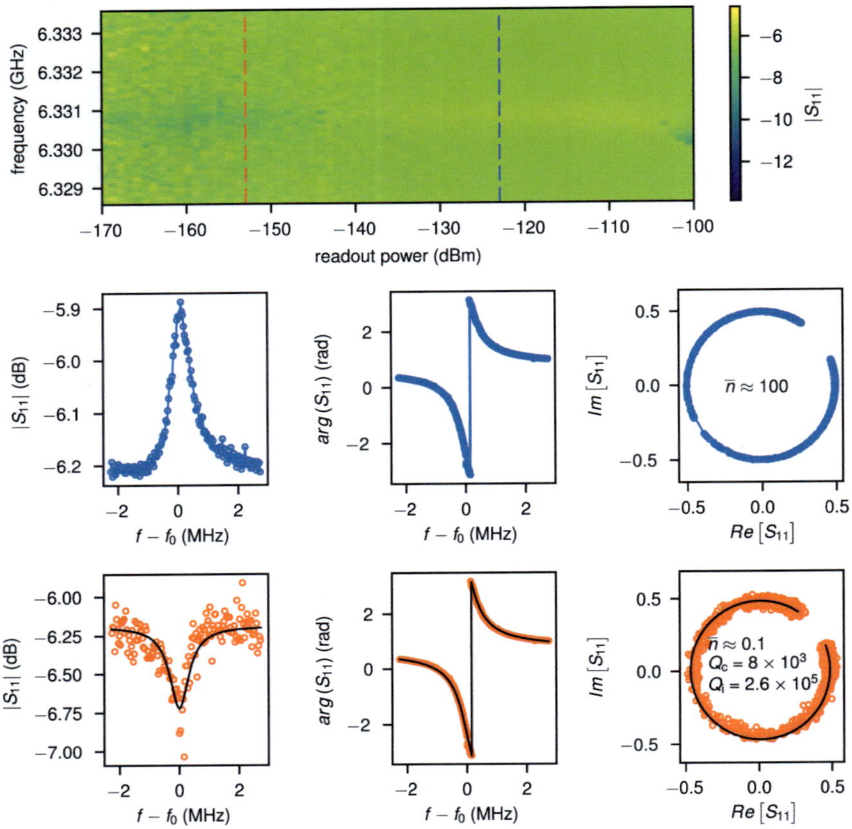

Figure 3.13: Measured data of resonator C2 with a resonant frequency of 6.3308 GHz. The amplitude response changes from a peak at high powers (blue dashed line in color plot and center row of panels) to a dip at low powers corresponding to $\overline{n} \approx 0.1$ (orange dashed line and bottom row of panels). From a fit to the low power data, we extract $Q_\mathrm{i} = 2.6 \times 10^5$ and $Q_\mathrm{c} = 8 \times 10^3$. Assuming a constant coupling, we estimate ~ 100 photons on average circulating in the resonator for the data indicated by the blue dashed line in the color plot of the raw amplitude data (for clarity, the points are joined by straight line segments).

3.2.2 MKID geometry resonators

The following section provides an overview of the results obtained from measurements of MKID type resonator samples. We group the results of all resonators measured on the same chip and show the median internal quality factor as a function of the average number of photons, while we also indicate the observed spread. Resonant frequencies, coupling quality factors and self-Kerr coefficients are given in a table for each set of resonators. In the following we present results of three chips, which were designed in the group of A. Monfardini at Institut Néel in Grenoble (France) and fabricated at KIT (cf. Fig. 3.3). The resonators are patterned by the previously mentioned *e*-beam lift-off lithography process of grAl film deposited in the Plassys™ *e*-beam evaporation system. A ground plane, enclosing all resonators, and the CPW transmission line is fabricated from a 20 nm thick aluminum film using optical lift-off lithography.

Sample M1: $R_{\square} = 1.0\,\mathrm{k}\Omega/\square$, $t = 20\,\mathrm{nm}$

Out of the 22 resonators patterned on chip M1, the raw data of 15 could be evaluated using the fit routine developed in Ref. [88] for a notch type coupling geometry. The frequencies of the resonators are distributed between 2.1284 GHz and 2.7356 GHz and are given in detail in Table 3.3 together with the mean coupling quality factor of each resonator and the self-Kerr coefficient extracted from a linear fit to the frequency as a function of the average number of photons in the resonator. By comparing the measured resonant frequencies with 2.5D FEM simulations, the kinetic inductance in these resonators is estimated to be $L_{\mathrm{kin}} = 0.6\,\mathrm{nH}/\square$.

Figure 3.14 gives an overview over the extracted internal quality factors of the 15 resonators. The black solid indicates the median Q_{i}, while the blue shaded area indicates the spread between highest and lowest measured Q_{i} at the respective average number of circulating photons. Due to their weak coupling and small line width, the SNR of the measurement was only sufficient to reliably extract fitting parameters down to $\overline{n} \approx 60$, where the median yields an internal quality factor on the order of 4×10^5. Similarly to the measurements of the rectangular strip resonators, we estimate $\overline{n}$ to be correct within an order of magnitude and correspondingly the extracted self-Kerr coefficients on the order of $K_{11} = 0.02\,\mathrm{Hz}$ are expected to be an order of magnitude result.

Table 3.3: Resonant frequencies, mean coupling quality factor and fitted self-Kerr coefficient of all resonators on sample M1.

f_0 (GHz)	2.1284	2.1868	2.2239	2.2883	2.3114	2.3374	2.4098	2.4519
$\overline{Q_{\mathrm{c}}}(\times 10^5)$	2.0	1.8	1.7	1.9	1.7	1.8	1.6	1.6
K_{11} (Hz)	0.02	0.02	0.03	0.02	0.03	0.02	0.02	0.03

f_0 (GHz)	2.4768	2.5018	2.5501	2.6002	2.6666	2.7215	2.7356
$\overline{Q_{\mathrm{c}}}(\times 10^5)$	2.2	1.8	2.0	2.1	2.3	1.9	2.1
K_{11}(Hz)	0.02	0.02	0.02	0.02	0.04	0.03	0.04

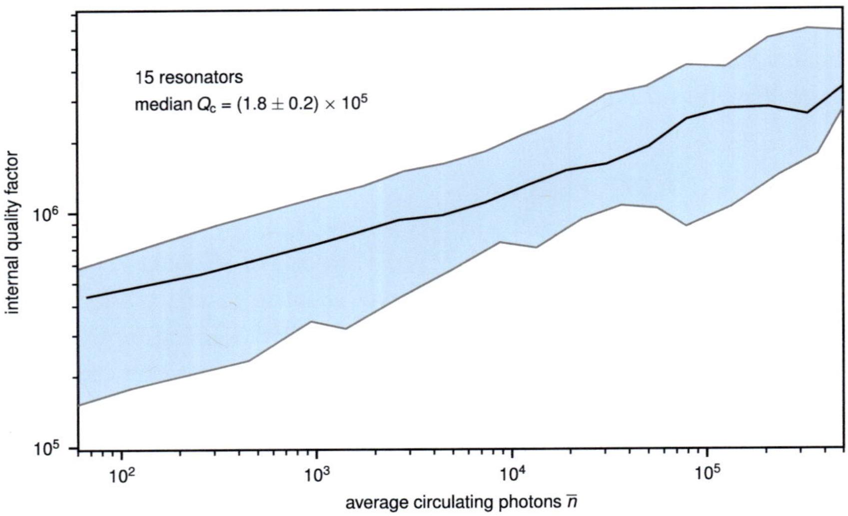

Figure 3.14: Median Q_i (black line) and range of extracted internal quality factors (blue shaded region) of 15 resonators on sample M1. At $\overline{n} \approx 60$ we measure internal quality factors ranging from 1.5 to 6×10^5 at median coupling quality factors $\overline{Q_\text{c}} = 1.8 \times 10^5$.

Sample M2: $R_\square = 1.0\,\text{k}\Omega/\square$, $t = 30\,\text{nm}$

Sample M2 was fabricated using the same layout for the resonators and an identical fabrication process to that of sample M1 presented in the previous section. In the measurement 20 resonators could be identified and fitted with frequencies between 2.1668 GHz and 2.9834 GHz. Table 3.4 lists the extracted resonant frequencies, mean coupling quality factors and extracted self-Kerr coefficients for all resonators. Where no self-Kerr coefficient is listed in the table, the frequency change of the resonator was not large enough over the measured power range to provide a basis for the linear fit. From successful linear fits to the change of the resonant frequency, we extract a self-Kerr coefficient on the order of 0.01 Hz. Since the measured resonant frequencies are similar to those of sample M1, we also estimate a kinetic inductance $L_\text{kin} = 0.6\,\text{nH}/\square$ for sample M2. Considering $L_\text{kin} \propto R_\square$ following Mattis-Bardeen formula, this is also expected.

Figure 3.15 shows the median internal quality factor of all 20 resonators as a function of $\overline{n}$ (black line). The shaded area indicates the range between the highest and lowest observed internal quality factors in the entire set of resonators. Again, poor SNR only allows to extract internal quality factors for average circulating photon numbers $\gtrsim 70$. In this photon number regime the median internal quality factor is on the order of 4×10^5, comparable to sample M1 (cf. Sec. 3.2.2).

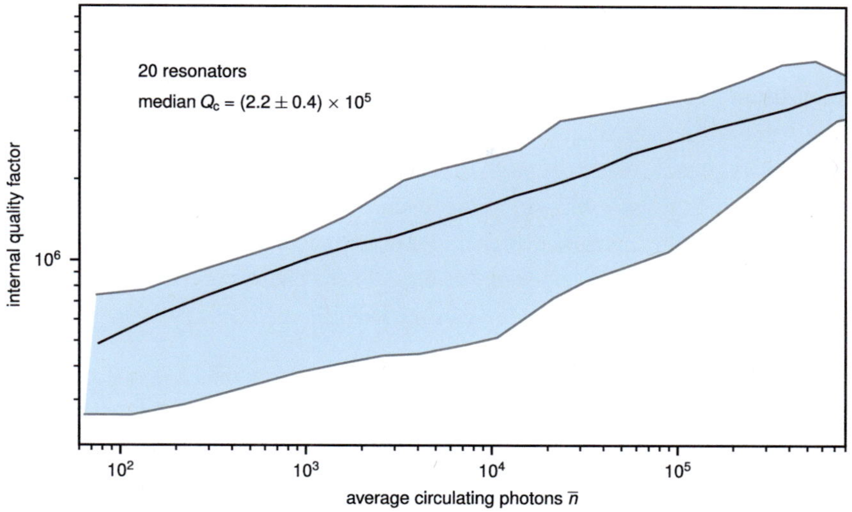

Figure 3.15: Median Q_i (black line) and range of extracted internal quality factors (blue shaded region) of 20 resonators on sample M2. At $\overline{n} \approx 70$ we measure internal quality factors ranging from 1.5 to 6×10^5 at median coupling quality factors $\overline{Q_c} = 2.2 \times 10^5$.

Table 3.4: Resonant frequencies, mean coupling quality factor and fitted self-Kerr coefficient of all resonators on sample M2. For resonators where no K_{11} is shown, the frequency shift as a function of readout power did not suffice to extract K_{11} from a linear fit.

f_0 (GHz)	2.1668	2.2037	2.2037	2.2215	2.2379	2.3185	2.3231	2.3623
$\overline{Q_c}(\times 10^5)$	2.6	2.2	2.2	2.4	2.3	2.2	2.8	2.0
K_{11} (Hz)	0.01	-	0.06	-	-	0.01	-	-
f_0 (GHz)	2.3826	2.4012	2.4604	2.5033	2.5099	2.5392	2.5674	2.6872
$\overline{Q_c}(\times 10^5)$	1.8	1.9	2.9	2.1	2.4	2.2	2.1	2.9
K_{11}(Hz)	-	0.01	0.01	-	0.01	0.01	-	-
f_0 (GHz)	2.7518	2.7836	2.7858	2.9834				
$\overline{Q_c}(\times 10^5)$	2.4	2.1	1.9	1.8				
K_{11}(Hz)	0.01	0.01	0.01	0.01				

Sample M3: $R_\square = 0.45\,\mathrm{k}\Omega$, $t = 20\,\mathrm{nm}$

In the measurements of the final sample of the MKID geometry M3, we could identify and fit 16 resonators with frequencies between 2.7644 GHz and 3.7238 GHz. Table 3.5 gives resonant frequency, mean coupling quality factor and fitted self-Kerr coefficient of all resonators. Similar to sample M2, for some resonators the self-Kerr could not be extracted due to a too small frequency shift, which could not be resolved. For the samples where K_{11} could be measured, we extract values on the order of 0.1 Hz. A comparison with 2.5D FEM simulations yields a kinetic inductance $L_{\mathrm{kin}} = 0.3\,\mathrm{nH}/\square$.

Table 3.5: Resonant frequencies, mean coupling quality factor and fitted self-Kerr coefficient of all resonators on sample M3. For resonators, where no K_{11} is shown the frequency shift as a function of readout power did not suffice for the linear fit to extract K_{11}.

f_0 (GHz)	2.7644	2.8391	2.8675	2.9474	2.9740	3.0395	3.0917	3.1517
$\overline{Q_c}(\times 10^5)$	2.4	1.3	3.5	1.1	1.2	0.8	1.0	1.6
K_{11} (Hz)	0.07	0.04	0.05	-	-	0.03	-	0.08

f_0 (GHz)	3.1523	3.2706	3.4046	3.4287	3.5485	3.6305	3.6509	3.7238
$\overline{Q_c}(\times 10^5)$	1.2	1.3	1.7	1.5	2.9	4.3	1.9	3.5
K_{11} (Hz)	0.06	0.04	0.09	0.07	0.09	0.10	-	0.25

Due to some stronger coupled resonators on sample M3, internal quality factors in the single photon regime could be extracted. Figure 3.16 shows the median Q_i as a function of the average number of photons circulating in the resonator (black line). The blue shaded area indicates the highest and lowest Q_i for the corresponding photon number. In the single photon regime, we extract a median Q_i on the order of 4×10^5.

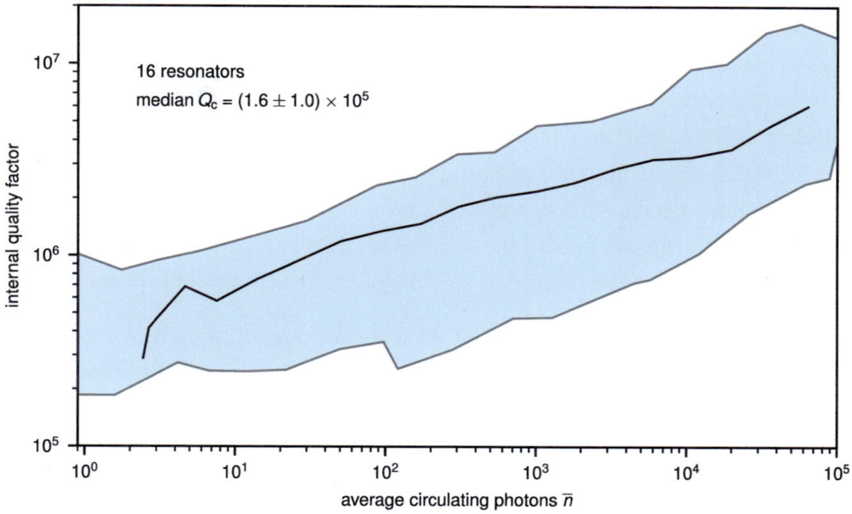

Figure 3.16: Median Q_i (black line) and range of extracted internal quality factors (blue shaded region) of 16 resonators on sample M3. In the single photon regime we measure internal quality factors ranging from 2 to 10×10^5 at median coupling quality factors $\overline{Q_\mathrm{c}} = 1.6 \times 10^5$.

3.3 Dielectric loss in grAl resonators

Following the encouraging results of the previous section, we investigate the limiting loss mechanisms in the measured grAl resonator samples. Using these insights, we work towards optimized design approaches for the grAl superinductor and try to develop mitigation strategies for the identified loss mechanisms. A common source of loss in superconducting circuits is dielectric, also referred to as capacitive, loss [143] whose mitigation by optimized designs, materials, and fabrication techniques is largely responsible for increasing coherence times in superconducting circuits [2]. Another usual suspect, especially at microwave frequencies, when due to the kinetic inductance of the superconductor the Cooper-pair condensate does not provide a perfect shunt, are quasiparticles. Both loss mechanisms have been shown to saturate with increasing readout power [42, 104, 144, 145] and could therefore be responsible for the observed increase of Q_i as a function of $\bar{n}$ (cf. Sec. 3.2).

However, it is not straightforward to unambiguously determine which of the loss mechanisms poses a limit to the observed coherence of a superconducting circuit. In this thesis we rely on a participation ratio based approach to investigate dielectric loss, which is based on FEM simulations of the sample geometry. The approach is outlined in Sec. 1.4.2 and further detailed in Refs. [90, 91]. To find the surface participation ratio, we model the measured resonator and its surrounding electro-magnetic environment as precisely as possible in a 3D FEM high frequency software (ANSYS HFSS™). The simulation does not only give access to the resonant frequency and quality factor of a mode of the modeled structure, but also provides the electric field distribution. Since dielectric loss channels mostly couple to the electric field, we numerically integrate the electric field distribution of the resonator in an area covering the resonator and extending to areas with significant electric field strengths in close vicinity to the resonator. Following the technique of Refs. [90, 91], we assume a thickness of 3 nm for the interface layer in which we integrate the energy stored in the electric field. Using this approach we extract the sum of metal-substrate, metal-air, and substrate-air participation ratios, which we refer to as *surface* participation ratio in short (cf. Fig. 1.14).

After obtaining the participation ratios of the measured resonators we plot the extracted internal quality factor in the single photon regime as a function of the surface participation ratio following the approach of Wang *et al.* [91]. Figure 3.17 shows the extracted single photon Q_i as a function of the surface participation ratio of resonators A2, B2, C2, MKID sample M3, resonators CPW1 to CPW3, and the data of one control experiment of pure aluminum resonators in the same 3D waveguide sample holder used for resonators A2 to C2 (cf. Sec. 4.1.3).

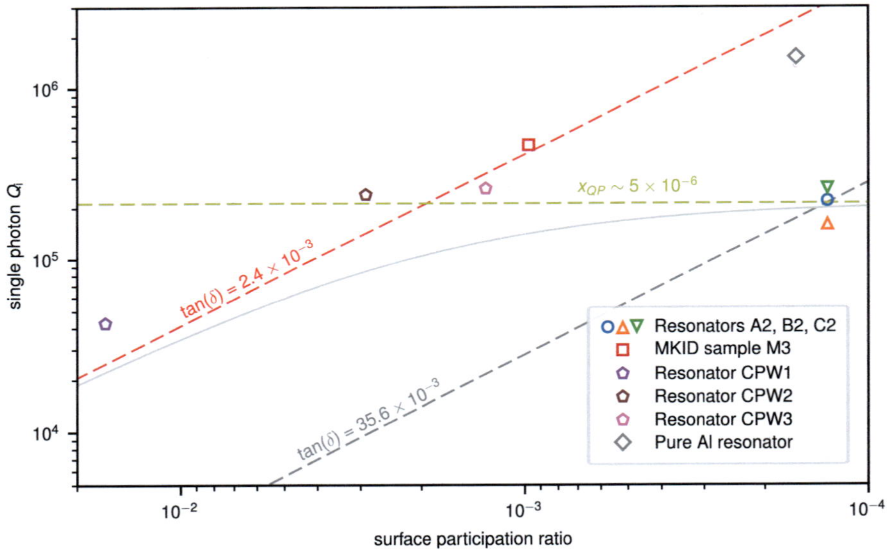

Figure 3.17: Comparison between single photon Q_i as a function of the surface participation ratio of different resonator geometries. The data suggests that resonators A2-C2 (orange square, blue circle, and green triangle) are not limited by dielectric interface loss, but by excess quasiparticles. In contrast, for resonators with surface participation ratio $\gtrsim 10^{-3}$ and pure aluminum resonators (grey diamond), we observe a dielectric loss tangent similar to previously reported values [91, 93]. For a detailed discussion see the main text.

Starting with the measurements on samples having the smallest surface participation ratio (A2, B2, C2) we would expect to observe the highest internal quality factor if dielectric loss is the limiting loss mechanism. Based on an expected interface dielectric loss of $\tan(\delta) = 2.6 \times 10^{-3}$ [91], we estimate the limit of Q_i imposed by dielectric loss $\sim 2 \times 10^6$ (cf. red dashed line). As can be seen in Fig. 3.17, we observe internal quality factors about 10 times smaller than estimated for resonators A2, B2, and C2 (cf. blue circle, orange triangle, and green triangle). This could be due to a ten times higher loss tangent either associated with our fabrication process or the grAl film, or the presence of an additional limiting loss mechanism. In principle, dissipation in the normal conducting copper waveguide could provide an additional loss channel, limiting the internal quality factor of resonators in this sample holder geometry to $\sim 2 \times 10^5$. From a measurement of resonators fabricated from pure aluminum (cf. Sec. 4.1.3) where we observe internal quality factors in the range of 10^6 (cf. grey diamond) in an identical copper waveguide sample holder we conclude that the waveguide is not limiting $Q_\mathrm{i} \lesssim 10^6$.

Control experiments of resonators with increasing participation ratio fabricated from grAl show similar internal quality factors for a ten times larger surface participation ratio (cf. CPW2 - brown pentagon, and CPW3 - pink pentagon). Increasing the participation ratio in CPW1 (violet pentagon) by ~ 2 orders of magnitude compared to resonators A2, B2, and C2, leads to a Q_i about a factor of 5 lower. From the internal quality factors of resonators CPW1-3 and their respective surface participation ratios we extract a dielectric interface loss tangent $\tan(\delta) = 2.4 \times 10^{-3}$ (cf. Fig. 3.17, red dashed line), which is in good agreement with previously reported values for pure Aluminum in various geometries and different laboratories [91, 93]. Therefore, we conclude that rectangular strip resonators A2-C2 are not limited by a particularly lossy dielectric interface layer.

Based on the observed sudden drops of the resonant frequency in resonator A2 (cf. Fig. 3.8) we suspect excess quasiparticles as the limiting loss mechanism in resonators A2-C2. A further hint in this direction is provided by the single photon Q_i of sample M3, which is higher despite a surface participation ratio approximately a factor of 10 larger. Due to the pure aluminum, lower superconducting gap ground plane enclosing the MKID type resonators, we attribute the higher Q_i to phonon trapping in the ground plane, which reduces the excess quasiparticle density produced by phonons with energy larger than twice the superconducting gap of grAl. In MKIDs the efficiency of reducing the quasiparticle creation by a lower gap material for phonon trapping has recently been shown and could have potential applications in superconducting quantum circuits [146, 147]. In the following section we investigate quasiparticle dynamics in the rectangular strip grAl resonators A2-C2 and identify them as a probable limiting loss mechanism.

3.4 Quasiparticle dynamics in grAl resonators

Two main reasons prompt the investigation of quasiparticle dynamics in granular aluminum resonators. First, following the investigation of the internal quality factor as a function of the surface participation ratio suggests that rectangular strip resonators in the 3D waveguide sample holder are not limited by dielectric loss. An alternative loss channel limiting the coherence of superconducting quantum circuits is excess quasiparticles, as seen in many different groups [39, 42, 101, 104, 148]. Secondly, during power sweep measurements of sample A2 sudden drops of the resonant frequency were observed, which are reminiscent of high energy particle impacts in MKIDs [96]. In this chapter, we will describe the event detection, the fitting procedure used to extract characteristic timescales of these events, the statistics of their occurrence and finally a phenomenological model linking the observed quasiparticle dynamics with the previously observed increase of Q_i as a function of the applied readout power.

Figure 3.18 shows a typical continuous measurement of the resonant frequency of resonators A2 (top) and B2 (bottom), which reveals regular sudden drops of the phase signal, corresponding to a shift of the resonant frequency to lower values (cf. Fig 3.20). We attribute these signatures to high energy events, which directly or indirectly deposit sufficient energy in the superconducting film of the resonators to break a large number of Cooper pairs. Since the kinetic inductance is inversely proportional to the number of Cooper pairs, L_kin increases and leads to a shift in resonant frequency. This process constitutes the detection principle of MKIDs [99], which are today used for radio astronomy applications [100].

We perform measurements like the ones shown in Figure 3.18 on resonators A2-C2 at various readout powers for a total time of 45 min each. From the measured raw phase data we automatically identify traces with a quasiparticle producing impact event and subsequently confirm all preliminary events manually. Table 3.6 shows the number of identified impacts at the different readout powers for each resonator.

Before proceeding with further data analysis we attempt to elucidate the sources of the observed events. To this end, we first histogram the durations between two successive events, which is shown in Fig. 3.19(a). We find an exponential distribution for each of the three resonators and fit the cumulated data of the three histograms. This yields a characteristic rate of $1/20\,\mathrm{s}$ for the impacts. The exponential decay indicates a Poisson process as the source for the events. Therefore, the occurrence of the events should not be correlated. In the MKID community, these events are attributed to impacts of high energy particles in the substrate, where they produce ballistic phonons [96, 147] which subsequently distribute across the entire chip and have sufficient energy to break Cooper pairs in the superconducting resonators [96, 146].

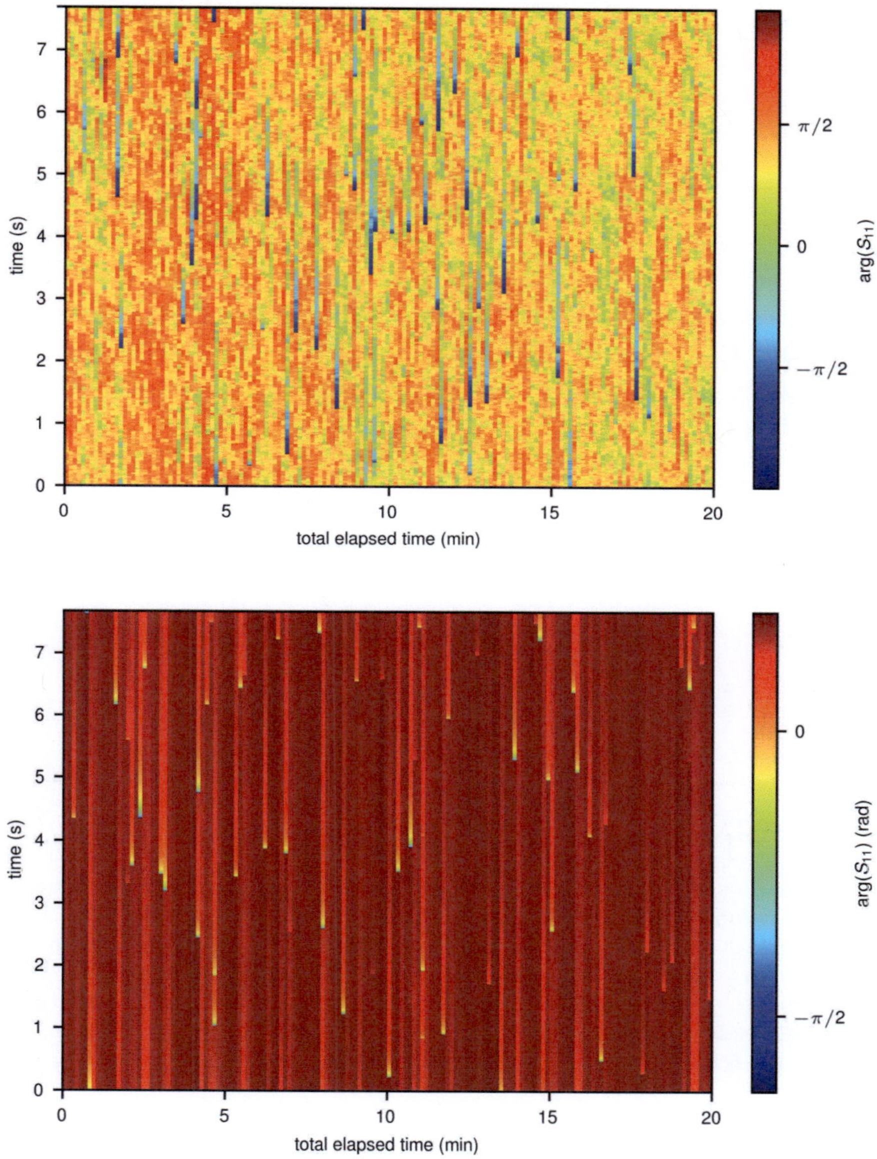

Figure 3.18: A quasi continuous monitoring of the resonant frequency of resonator A2 (top) and B2 (bottom) reveals regular stochastic sudden drops of the phase signal. By monitoring consecutive time traces of the resonator phase response, which are each $\sim 7\,\mathrm{s}$ long, we record a timespan of about 45 min (for clarity only partially shown). Due to its smaller line width, resonator A2 exhibits a larger sensitivity and smaller impacts are also visible.

Candidates for the source of the observed events are cosmic particles with high energies, which penetrate all shielding layers in our experiment, particles from radioactive decay, possibly inside the sample holder, or infrared photons, which are also suspected as a source of decoherence in superconducting qubits [149]. Figure 3.19(b) shows a histogram of the frequency shift of the first point of an event normalized to the corresponding resonant frequency of the resonator. For resonator A2 there is a clear peak observable, indicating that the source of the quasiparticle bursts has a certain energy. The histograms for resonators B2 and C2 are less clear. To draw more concrete conclusions about the origins of impacts, the histograms would need to be corrected for the different sensitivities of the three resonators to quasiparticle bursts. For a more detailed investigation of the origin of the observed impacts calibrating the response of grAl resonators to deposited energy using a known source would be advantageous. The calibration could potentially help to identify the energy range of the naturally observed impacts, and therefore narrow down the range of possible sources of quasiparticle bursts.

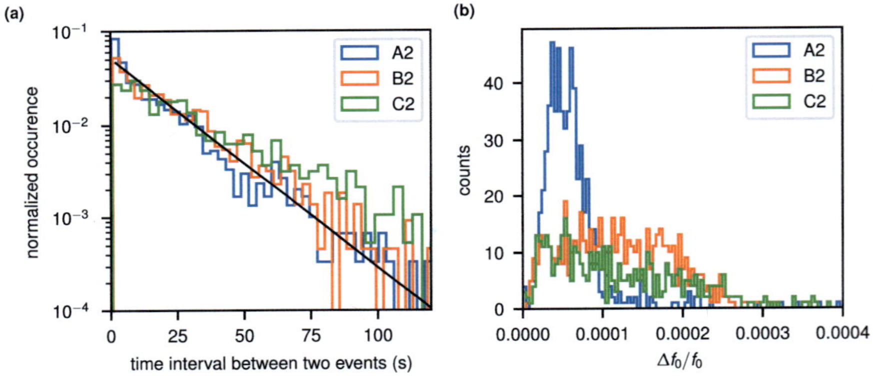

Figure 3.19: Histogram of the times between two successive impacts and histogram of normalized event amplitude for all available data (cf. Tab. 3.6). **(a)** We find an exponential distribution of the intervals between two quasiparticle events. This distribution suggests a poissonian process producing the observed quasiparticle impacts. From the exponential fit to the cumulated histograms, we extract an average rate of $1/20\,\mathrm{s}$ for the impacts. **(b)** Histogram of event amplitude in frequency normalized to the corresponding resonant frequency. For resonator A2, which is the most sensitive resonator due to the smallest line width we observe a peaked distribution. Resonators B2 and C2 show broader distributions with no clearly distinguishable peak.

The individual quasiparticle events consist of a very steep drop of the resonant frequency, which is much faster than resolvable with the vector network analyzer, followed by a relaxation process over a timescale of seconds. Similar to what has been observed in superconducting transmon qubits [43], we regard the relaxation process as a combination of two physical processes each with a characteristic timescale. Initially, the relaxation is governed by the recombination of quasiparticles into Cooper pairs with a characteristic timescale that is presumably strongly correlated with the electron-phonon interaction in grAl. In this process we believe the dynamics to be a function of the number of generated quasiparticles, their initial distribution, and background quasiparticle density. Therefore, we conclude that the first steep decay is non-uniform between the different observed impacts. After the primary relaxation process dominated by recombination, during which the excess quasiparticle density is considerably reduced, the dynamics is presumably governed by a trapping or diffusion process whose characteristic timescale is proportional to the number of quasiparticles in contrast to the recombination process, which is proportional to the square of the number [43]. Therefore, we observe an exponential decay of the excess quasiparticle density back towards the equilibrium value. From our measured data, we believe this exponential part of the relaxation process to be universal for a given resonator at a given readout power.

Table 3.6: Number of individual events identified at the different readout powers during a total measurement time of 45 min. First, we identify traces with a QP burst automatically, followed by a manual check of all identified traces.

readout power	number of identified events		
(dBm)	A2	B2	C2
-120	103	93	72
-125	84	87	56
-130	75	79	54
-135	58	79	63
-140	72	81	62
-145	54	89	56
-150	59	62	56
-155	47	23	30
-160	48	-	-
-165	39	-	-

Figure 3.21 illustrates the joint fitting procedure for the exponential relaxation of all events of a dataset. First, we shift the individual events with respect to each other such that the overlap of the exponential tails is maximized [see Fig. 3.21(a)]. After performing this for all events, where the initial steep part of the relaxation process is removed from all but the reference event, we average the remaining data to generate the mean exponential decay curve [cf. Fig. 3.21(b)]. As can be seen in the panel the noise increases for longer times, as there are fewer events available for the averaging and the deviation from the equilibrium value is exponentially decreasing. Finally, we manually identify the exponential part of the averaged relaxation curve [cf. black dashed lines Fig. 3.21(c)], which we use to fit an exponential function and extract the characteristic relaxation time τ_{ss}.

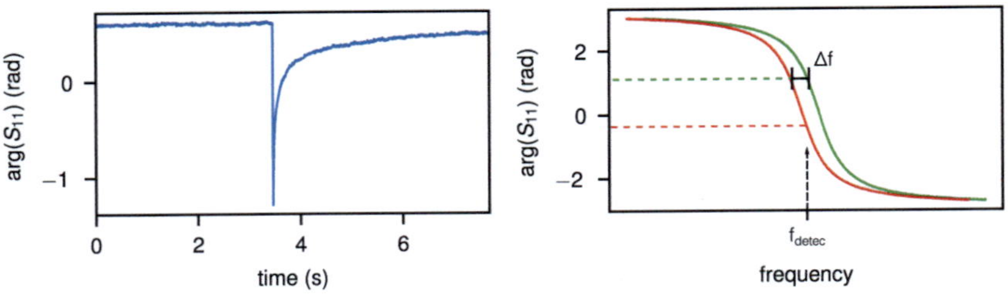

Figure 3.20: Line plot of a single time trace with an impact recorded in resonator B2 and schematic illustration of the detection principle. The left panel illustrates the steep initial drop of the phase signal, indicating a drop in resonant frequency, followed by a slow relaxation back to the equilibrium value over a timescale of $\sim 1\,\mathrm{s}$. This relaxation time is orders of magnitude slower than observed in pure aluminum MKIDs [96]. The right panel schematically shows the detection principle, which is similar to a standard circuit quantum electrodynamics (cQED) [50] readout. A quasiparticle impact will shift the resonator from its equilibrium position (green curve) to a lower frequency (red curve), which we observe as a drop in the phase signal at the detection frequency f_{detec}. Inverting the frequency-phase spectrum allows to calculate the frequency shift Δf.

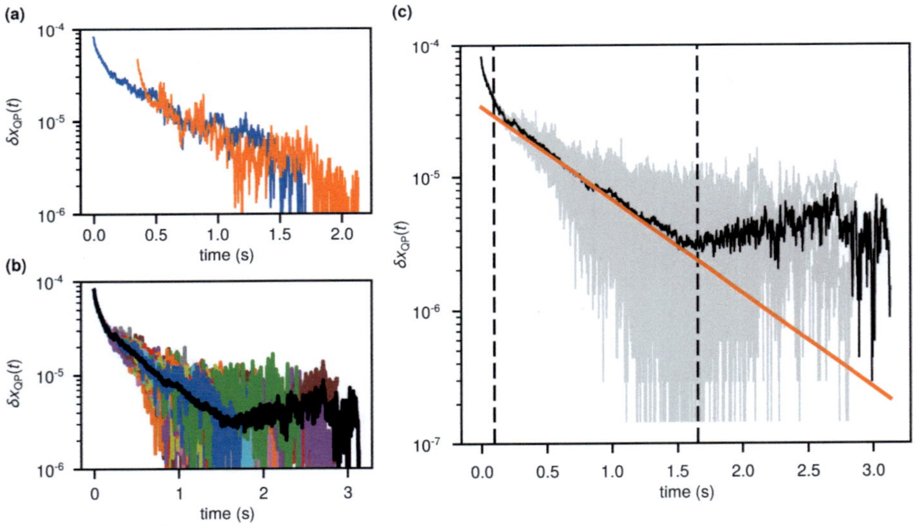

Figure 3.21: Fit procedure of combined events to extract the exponential tail's time constant. **(a)** After automatically identifying and manually checking the events of a dataset in the phase signal, we convert the phase data to the relative frequency shift of the resonator $\delta f/f_0$. This relative frequency shift is proportional to the normalized excess quasiparticle density $\delta x_{\mathrm{QP}} \propto 2\delta f/f_0$ [43]. For further processing, we shift the events such that we achieve maximal overlap of the tails, which amounts to a rescaling of the amplitude of the exponential tail, not changing the time constant. **(b)** From the dataset of all shifted traces, we find an average event (black line). **(c)** In a final step, we identify the exponential part of the averaged relaxation trace (dashed black lines) and fit an exponential decay to obtain the relaxation constant τ_{ss}.

Figure 3.22 shows the time constant of the exponential decay τ_{ss} as a function of the average number of circulating photons and its correlation with the fitted internal quality factor for resonators A2, B2, and C2. The left panel shows a decrease of the relaxation time with an increasing number of photons for the smallest resonator A2 (orange squares) from 1 s to 7 s at low photon numbers to ~ 0.6 s at the highest readout power. The observed scatter between the points could potentially be caused by a fluctuating background quasiparticle density, as observed in superconducting fluxonium qubits [16, 101]. In resonators B2 and C2 with 2, respectively 7 times larger width, we observe τ_{ss} constant within $\pm 50\%$ across the range of investigated powers. The orange line is a fit to a phenomenological model, which links τ_{ss} with the activation of localized quasiparticles by photons and is discussed in more detail further below. Visible in the right panel of Fig. 3.22, $Q_i(\bar{n})$ and $\tau_{ss}(\bar{n})$ appear to be correlated for resonator A2 (orange squares), whereas no correlation is visible for resonator B2 (blue circles).

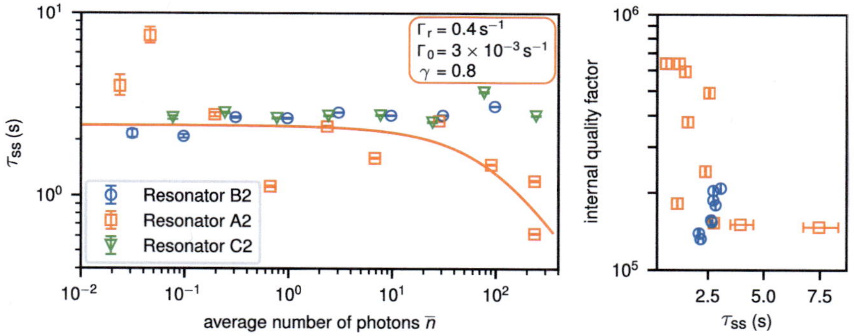

Figure 3.22: Exponential relaxation constant τ_{ss} as a function of the average number of photons in the resonator and correlation between τ_{ss} and Q_i. For resonator A2, which has the smallest dimensions, we observe a decrease of the relaxation time constant at higher circulating photon numbers. The wider resonators B2 and C2 show within a factor of two a time, which is constant with readout power. In the right panel we show the correlation between internal quality factor and τ_{ss}. A slight correlation can be observed for resonator A2.

Based on the results presented in Sec. 3.3, we concluded that grAl resonators A2, B2, and C2, are not limited by dielectric interface loss. Since they show pronounced events that shift the resonant frequency to lower values, a signature of an increase in the excess quasiparticle density, we investigate this as a limiting loss mechanism. Furthermore, Fig. 3.22 suggests a correlation between $Q_i(\bar{n})$ and $\tau_{ss}(\bar{n})$ for resonator A2 (orange squares). Based on recent theoretical results [150], and thanks to a close collaboration with Gianluigi Catelani from FZ Jülich, we propose a phenomenological model linking quasiparticle dynamics in grAl resonators with $\bar{n}$, and with the observed increase in the internal quality factor [46] (cf. Sec. 3.2). It has been previously reported that circulating power can accelerate QP diffusion, for example in qubits [104] and small Josephson junctions [42]. Disorder in grAl films could lead to spatial variations of the superconducting order parameter, resulting in a small sub gap tail of the density of states (DOS) where quasiparticles could localize [150]. Similar to the approach of Rothwarf and Taylor [151] we propose a phenomenological model attempting to mimic the observed quasiparticle dynamics. In the model we assume a localized x_l and mobile x_m quasiparticle density governed by the following rate equations

$$\dot{x}_m = -\Gamma_{mm}x_m^2 - \Gamma_{ml}x_m x_l - \Gamma_{\text{loc}}x_m + \Gamma_{\text{ex}}x_l + g_m, \tag{3.1}$$

$$\dot{x}_l = -\Gamma_{ll}x_l^2 - \Gamma_{ml}x_m x_l + \Gamma_{\text{loc}}x_m - \Gamma_{\text{ex}}x_l + g_l. \tag{3.2}$$

Here, Γ_{mm}, Γ_{ll}, and Γ_{ml} describe the recombination rate of two mobile, two localized, or a mobile and a localized quasiparticle. Γ_{loc} and Γ_{ex} are rates of quasiparticle localization in, and excitation out of states in the sub gap DOS, and g_m and g_l represent the generation of mobile and localized quasiparticles. In the model, the photon assisted excitation of localized quasiparticles is expressed by setting $\Gamma_{\text{ex}} = \Gamma_0 \bar{n}$, where Γ_0 accounts for the interaction strength. Simplifying the model (see Appendix B) yields an expression for the dependence of τ_{ss} on the average number of circulating photons in the resonator,

$$\frac{1}{\tau_{\text{ss}}} = \Gamma_r + \Gamma_0 \left[\bar{n} + \frac{1}{2\gamma}\left(\sqrt{1+4\gamma\bar{n}} - 1\right)\right]. \tag{3.3}$$

The exponential relaxation rate consist of a residual, constant rate Γ_r, and a part $\propto \Gamma_0 \bar{n}$. In the photon number dependent part $\gamma = 2\Gamma_{\text{loc}}\Gamma_0/(g_m\Gamma_{ml})$. In similar fashion we also obtain an expression for the loss caused by localized quasiparticles, which is also photon number dependent now,

$$\frac{1}{Q_i} = \frac{1}{Q_0} + \beta \left[\frac{1}{1 + \frac{\gamma\bar{n}}{1+\frac{1}{2}(\sqrt{1+4\gamma\bar{n}}-1)}} - 1\right]. \tag{3.4}$$

At low numbers, the internal quality factor reaches a residual value Q_0 independent of the photon number $\bar{n}$. In Eq. (3.4) $\beta \propto \Gamma_{\text{loc}}/\Gamma_{ml}$ and also takes into account quasiparticle-photon coupling strength and the DOS.

The solid orange line in Fig. 3.22 shows a fit of Eq. 3.3 to the data of resonator A2. As can be seen, the model captures well the observed trend in $\tau_{\text{ss}}(\bar{n})$, with fit parameters $\Gamma_r = 0.4\,\text{s}^{-1}$, $\Gamma_0 = 3\times10^{-3}\,\text{s}^{-1}$, and $\gamma = 0.8$. However, due to the large scatter in τ_{ss} and only a single resonator which shows a change in $\tau_{\text{ss}}(\bar{n})$ a more quantitative analysis of the phenomenological model needs further, more detailed experiments.

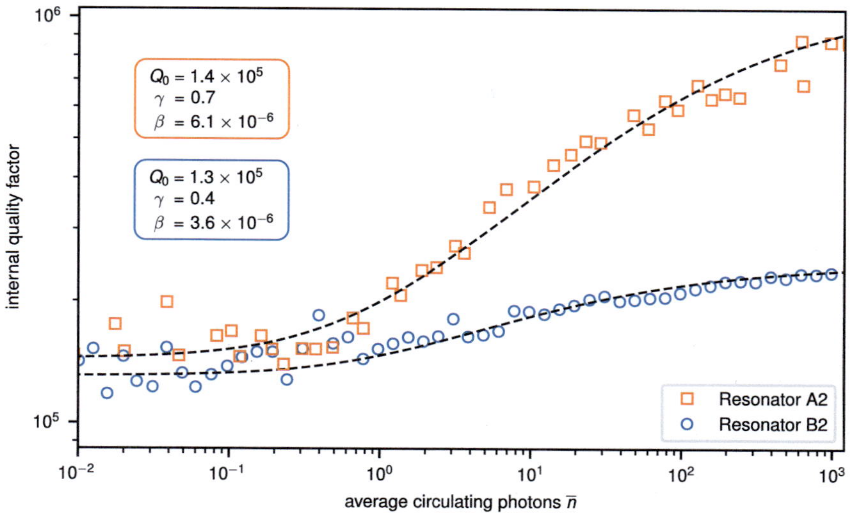

Figure 3.23: Internal quality factor as a function of $\bar{n}$ for resonators A2 and B2 fitted with a phenomenological model attributing the increase of Q_i to the activation of localized quasiparticles. The fit of the phenomenological model captures the measured data well.

Figure 3.23 shows the internal quality factor Q_i as a function of the applied readout power for resonators A2 and B2. The dashed black lines are fits of Eq. (3.4) to the measured data, with the fit parameters given in the correspondingly colored boxes in the plot. Both resonators show comparable residual quality factors on the order of 1×10^5. The coefficients γ and β differ approximately by a factor of 2 between resonators A2 and B2 (cf. correspondingly colored boxes in Frig. 3.23). A possible explanation is the smaller cross section of resonator A2, which could lead to a stronger quasiparticle-photon coupling, which influences the coefficients γ and β. However, since τ_{ss} of resonator B2 is flat across the measured range of average circulating photon numbers, the observed increase of Q_i could also be attributed to the suppression of residual dielectric loss. In conclusion, for a more quantitative insight into the dynamics of quasiparticles in rectangular strip grAl resonators further experiments exploring a larger parameter space of geometries, possibly with an optimized measurement setup for the time resolved recording of the resonant frequency are necessary.

As a final step we perform similar measurements as in Fig. 3.18 at a temperature of 200 mK. Figure 3.24 shows exemplary events recorded in all three resonators. At these temperatures, due to thermal excitation, the background quasiparticle density is increased, which leads to a significant speedup for the relaxation time of the observed quasiparticle events. Compared to measurements at the base temperature of our cryostat (25 mK), the relaxation process is ~ 2 orders of magnitude faster, and not resolvable well enough anymore to perform a quantitative analysis similar to the data measured at base temperature. Nevertheless, these measurements further confirm quasiparticles as the source of the observed events and could also provide a starting point for further, more detailed investigation of quasiparticle dynamics in grAl resonators.

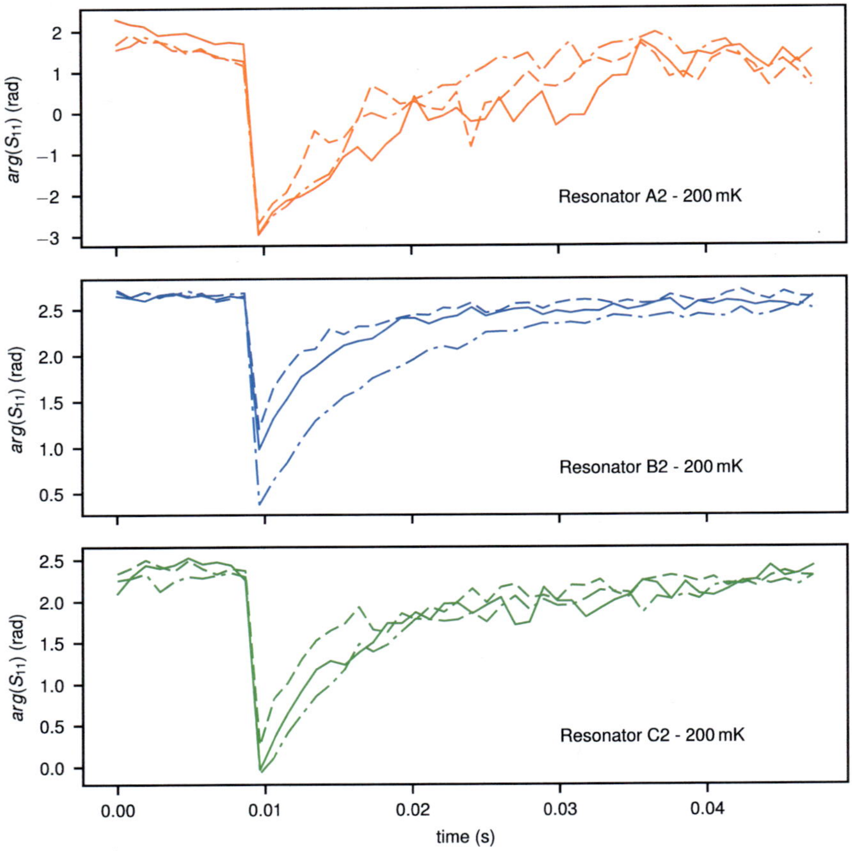

Figure 3.24: Impacts in rectangular strip resonators A2-C2 at 200 mK. Impacts similar to the ones observed at $T \approx 20\,\mathrm{mK}$ are also visible at higher temperatures. The relaxation time of the events is decreased by about two orders of magnitude compared to the low temperature results.

4 Granular aluminum fluxonium qubit

Building on the encouraging results of Chapter 3, which show the potential of granular aluminum for the realization of superinductors in quantum circuits, in this chapter we describe the design, fabrication, and measurement of a fluxonium qubit with a superinductor fabricated from a grAl thin film. As already discussed, the susceptibility of the fluxonium to different loss channels such as capacitive, inductive, or dissipation caused by quasiparticle tunneling across the circuit's small Josephson junction can be in-situ tuned by an external magentic flux (cf. Sec. 1.2). This feature makes the fluxonium an ideal testbed for the investigation of superinductors realized using grAl.

In the following chapter, we will discuss the design and fabrication of the grAl fluxonium, present spectroscopic measurements and finally conclude with a discussion of the measured coherence in the investigated samples.

4.1 Design and fabrication

Similarly to the rectangular strip resonator measurements, we also employ the 3D copper waveguide sample holder for our experiments involving the grAl fluxonium. The relatively straightforward design of the coupling between a resonator and the waveguide mode combined with the well controllable electro-magnetic environment, motivate the choice of sample holder geometry. For a discussion regarding the sample holder, which follows the approach reported in Ref. [102], see Sec. 3.1.1. A general description of the simulation technique for the coupling between the waveguide and a superconducting circuit can be found in Sec. 3.1.2.

In Sections 4.1.1 and 4.1.2 we describe the design process for the readout resonator and fluxonium, and give the chosen circuit parameters for the inductively coupled fluxonium-resonator system. Following that, we outline two fabrication approaches for the realization of a fluxonium qubit with grAl superinductor in Section 4.1.3. Both of these approaches allow the fabrication of the small Josephson junction and the superinductor in one lithography step. The first process using two angled evaporations leads to a design completely realized in grAl. In contrast, for the second technique we employ two angle evaporations of pure aluminum to form the Josephson junction and readout resonator, followed by a zero angle deposition of the grAl superinductor. We pursued both fabrication approaches. However, the latter first produced the working fluxonium samples which were measured during this thesis, and we did not continue fabrication in parallel with the measurements.

In the last part of Section 4.1.3 we describe a cleaning process to remove native aluminum oxide layers [95]. This cleaning process allows to coherently contact aluminum layers realized in different lithography steps, which allows to connect arbitrarily shaped grAl superinductors to Josephson junctions (cf. Ref. [152, 153]).

4.1.1 Readout resonator

Among the characteristics which make superconducting quantum circuits one of the most promising systems for the realization of a quantum computer is the straightforward readout technique of superconducting qubits via the dispersive readout scheme [49, 50]. Depending on its state, the qubit induces a frequency shift in the resonator, which is dispersively coupled to the qubit, meaning that the coupling strength between qubit and resonator g is much smaller than the frequency difference between both systems, the so called detuning Δ.

How fast information can be gathered from a resonator is proportional to the inverse of the resonators linewidth $1/\kappa$, however, since resonator and qubit are coupled, the qubit is connected to the environment through the resonator. Therefore, the so called Purcell effect (or radiative loss) potentially limits the excited state lifetime of a qubit [154]. Furthermore, as a rule of thumb the resonator linewidth κ should be comparable to the dispersive shift of the resonator χ, to achieve high readout fidelity. Following these considerations, we target a resonator linewidth $\sim$ MHz, which at resonant frequencies in the 6 GHz to 8 GHz range corresponds to total quality factors $Q_\mathrm{l} \sim 10^3$. Ideally, the total quality factor should be entirely limited by coupling to the measurement setup, such that we aim for a coupling quality factor $Q_\mathrm{l} \approx Q_\mathrm{c} \sim 10^3$. This implies $Q_\mathrm{i} \gg Q_\mathrm{c}$ at $\bar{n} \approx 1$, which means that much less photons are lost to intrinsic dissipation in the resonator compared to the number of photons that is transmitted to the measurement apparatus.

Figure 4.1 shows the readout resonator and fluxonium design as well as an electrical circuit diagram. Since the resonator couples via its dipole moment to the propagating TE_{10} mode of the waveguide and closely resembles a dipole antenna, we also refer to the readout resonator as the readout antenna or simply antenna. All inductive elements of the antenna are implemented using grAl wires, which allows to tune the inductance by simply adjusting the number of squares of a given inductor.

As a first constraint for the ensuing readout resonator geometry design, we chose the targeted grAl film resistance. As a conservative choice, we aim at sheet resistances of 300 $\Omega/\square$ and

600 Ω/□, which ensures grAl film resistivities significantly below 10^4 μΩ cm, where the material undergoes a superconductor-to-insulator transition (SIT) [82]. As discussed above we target a coupling quality factor $Q_c \sim 10^3$ at a resonant frequency $f_0 \sim 7$ GHz, which we achieve using the simulation technique outlined in Sec. 3.1.2. The resulting antennae have a total length of 1 mm and width of 10 μm, which is also comparable to the dimensions of resonator B2 with $Q_c \approx 4 \times 10^3$ (cf. Sec. 3.2.1). Table 4.1 lists the design values of four antennae to be patterned on the same chip, each with a nominally identical fluxonium qubit coupled to it.

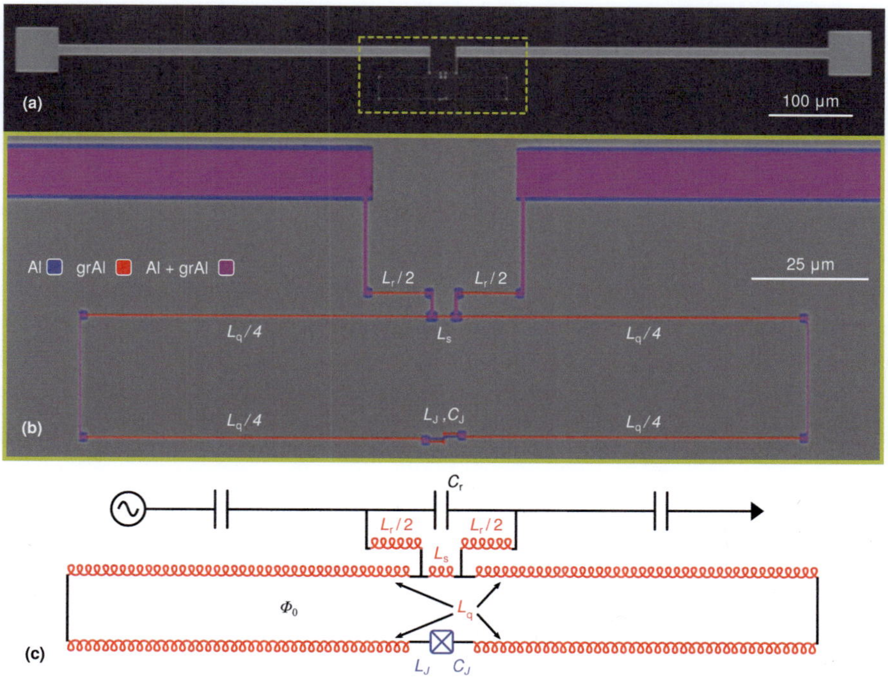

Figure 4.1: Optical images and schematic circuit of the granular aluminum fluxonium. **(a)** Overview of the entire circuit. The long wires are the readout resonator and provide the dipole moment for coupling to the waveguide. Two pads on both ends of the resonator are used for room temperature resistance measurements. **(b)** A zoom in [cf. yellow box in (a)] shows the fluxonium circuit. Granular aluminum wires (false colored in red) are used to implement all inductive elements of the circuit. The small Josephson junction, which is shunted by the grAl superinductor is fabricated with pure aluminum electrodes in a hybrid Niemeyer-Dolan-bridge free technique [59–61]. **(c)** Electrical schematic of the fluxonium resonator circuit. The equivalent circuit allows to connect the fluxonium circuit elements with the fabricated structures [cf. panel (b)].

Table 4.1: Design values for the readout resonators of grAl fluxoniums in a 3D rectangular waveguide sample holder.

L (nH)	C (fF)	resonant frequency (GHz)
21	22	7.405
23	22	7.075
25	22	6.786
27	22	6.530

4.1.2 Granular aluminum fluxonium

After converging on four sets of parameters for the readout resonators, the present section briefly discusses the design of the fluxonium qubits. To implement the coupling between qubit and resonator we use an inductive coupling scheme. This scheme has been successfully used before [16, 101, 102], and allows to harness the control over the shared inductance by adjusting the length of a grAl wire without the need of any further simulations to precisely design a coupling capacitance. Aiming for a large readout signal, we target a dispersive shift at the fluxonium half flux sweet spot comparable to the total linewidth of the readout resonator κ. Harnessing the capabilities of a numerical tool [51], which efficiently implements the Hamiltonian diagonalization approach presented in Ref. [30], we pick a set of fluxonium parameters according to the mentioned requirement $\chi \sim \kappa$. Finally, we converge on a fluxonium parameter set, that is similar to designs reported in the literature [16, 101, 102], see Tab. 4.2.

Table 4.2: Design values for the inductively coupled grAl fluxonium readout resonator system.

L_r	L_q	L_s	C_r	C_J	E_J/h
25 nH	375 nH	3 nH	22 fF	5 fF	11 GHz

A challenging task in the design of the fluxonium is to ensure a high enough frequency of the self-resonant modes of the superinductor, which should be well above the qubit operating frequency range of approximately $\sim 0.5\,\mathrm{GHz}$ to $12\,\mathrm{GHz}$. Extensive numerical simulations were carried out to fulfill this, and further constraints on the superinductor design, such as the total inductance (cf. Tab. 4.2), and superinductor wire geometry [51]. As a result of this simulation effort, a width for the superinductor wire of 120 nm was chosen, which allows to accommodate the desired inductance using a $R_\square = 300\,\Omega/\square$ grAl film, while limiting the added capacitance due to the wire, which ensured that the first self resonant mode of the superinductor is above $\sim 14\,\mathrm{GHz}$. The final fluxonium design for a targeted grAl sheet resistance $R_\square = 300\,\Omega/\square$ consists of a Josephson junction with an area of $0.1\,\mu\mathrm{m}^2$ shunted by a $\sim 150\,\mu m \times 24\,\mu m$ rectangular shaped superinductor loop, which is inductively coupled to the readout resonator via a 3 µm long shared grAl wire (see Fig. 4.1). On the same wafer, we also add a chip design for a $R_{\square} = 600\,\Omega/\square$ grAl film, by simply using grAl wires with half the length compared to the lower resistance design.

4.1.3 Superinductor implementation techniques

The following section outlines three approaches to connect a grAl superinductor to other parts of a quantum circuit. Two of the techniques employ a single lithography step fabrication process at the cost of some added design complexity. The third approach offers greater flexibility in terms of the circuit's geometry, but necessitates at least two lithography steps and a connection between lithography steps, which preserves the circuits quantum coherence. For connections to pure aluminum layers, the latter issue could be solved by an argon milling process, which removes the insulating AlO_x layer, that naturally forms on aluminum layers if exposed to oxygen, and allows coherent contacts (see Ref. [95], cf. Refs. [155, 156]).

Two angle approach

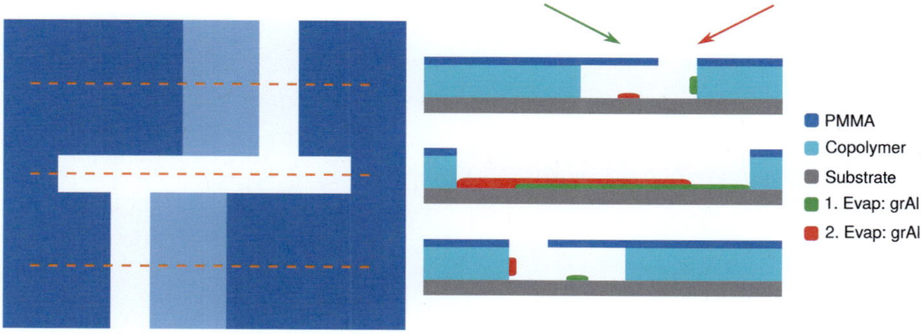

Figure 4.2: Circuit fabrication using the bridge-free technique for the Josephson junction [61]. The left panel shows the lithography mask for the small Josephson junction of the fluxonium whereas the right panel show cuts at the positions of the orange dashed lines. White areas indicate sections were both resists are completely removed. Light blue indicates an intact top layer resist, below which the bottom resist is removed to form the asymmetric undercut [61]. In the first evaporation (green), we angle deposit grAl, which forms the bottom Josephson junction electrode as well as its connecting wires, making up about half of the superinductor loop. After an oxidation forms the insulating barrier of the Josephson junction, we deposit grAl under the opposite angle (red), completing the Josephson junction and closing the superinductor.

The two angle approach relies on the fabrication of asymmetric undercuts to be able to build the Josephson junction using the bridge free technique [61]. As demonstrated already by L. Sun *et al.* [157] Josephson junctions with grAl electrodes show coherence on the order of µs, despite the disordered nature of the grAl electrodes. Fig. 4.2 schematically shows the junction area lithography mask employing the bridge-free technique described in Ref. [61]. For the design of the superinductor length and geometry one has to consider that wires parallel to the evaporation direction are deposited twice and consequently have approximately half the inductance. Furthermore, the superinductor loop is closed through a second junction, which is either the top or bottom parallel wire and should be designed large enough such that it does not add nonlinearity into the circuit. In case the readout antenna should not be fabricated from grAl, a second lithography step is necessary, which also needs a coherent contact between different lithographic layers (cf. Sec. 4.1.3).

Three angle approach

A benefit of the three angle process illustrated in Fig. 4.3 is the ability to implement the readout antenna for our current design from pure aluminum without any connecting layers or additional lithography steps. The first step is the fabrication of a conventional $Al/AlO_x/Al$ Josephson junction by two shadow evaporations and an oxidation step in between. In this step a Niemeyer-Dolan bridge as well as asymmetric undercuts are utilized. The superinductor wires are $\sim 120\,\mathrm{nm}$ wide and have no intentional undercut. Therefore, the angle deposited aluminum lands on the resist sidewalls and is removed during lift-off. This prevents a shunting of the grAl wires by the smaller impedance pure aluminum film. Structures such as the readout resonator, which are much wider than the superinductor wires allow the deposition of aluminum on the substrate, even though we employ a shallow angle of $\pm 30^\circ$ for the aluminum evaporation step. Additionally, grAl wires parallel to the evaporation direction of the aluminum (two short sides of the superinductor loop) are shunted. In the final fabrication step, we realize the grAl superinductor by a zero angle deposition.

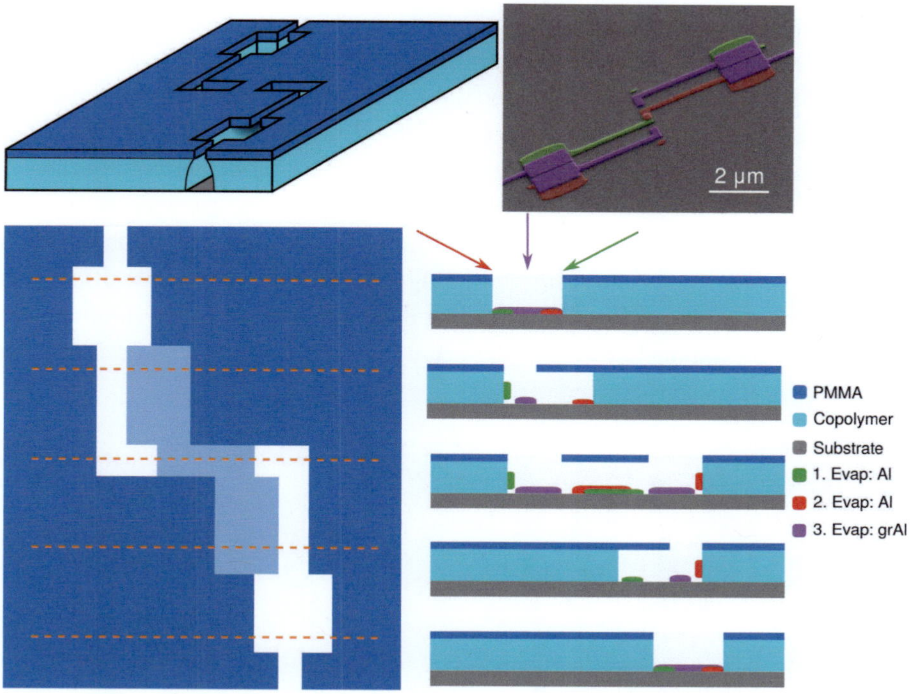

Figure 4.3: Three angle fluxonium fabrication. The dashed lines in the top view drawing indicate the location of the vertical cuts shown on the bottom right. Using the illustrated process allows the in-situ integration of $Al/AlO_x/Al$ Josephson junctions with a grAl superinductor in a single lithography step. First, we deposit two pure aluminum layers from opposite angles (indicated in green and red), interrupted by a static oxidation step, which forms the Josephson junction. In a third step, we deposit the grAl wires using a zero angle evaporation. Since we design the superinductor wires $\sim 120\,\mathrm{nm}$ wide, the aluminum layers land on the vertical resist walls and get lifted off. Only wider structures, for example the readout resonator antenna (cf. Fig. 4.1), allow for the deposition of pure aluminum, which shunts the grAl layer.

Argon ion milling for coherent superconducting contacts

The two step lithography approach in principle allows arbitrary geometries compared to the three angle approach for the grAl superinductor, since it is realized by a zero angle deposition. Josephson junctions can be added to the design as a first or second lithography step and can be implemented by any fabrication process. A key technology needed for multi-layer quantum circuits is a cleaning process of the native oxide on aluminum layers, which allows a good galvanic contact and ensures quantum coherence. In the following we describe an argon milling process, which provides both requirements at a level of $Q_\mathrm{i} \sim 10^6$ [95, 155, 156].

We perform the argon ion milling step using a Kaufman ion source [158], which is directly connected to the load lock of our Plassys MEB 550 S shadow evaporation machine. To remove the native aluminum oxide layer, we use 4 sccm of Ar gas, a beam voltage of 400 V, and an accelerating voltage of 90 V at a beam current of 15 mA. First, we calibrate the duration of the milling process by DC measurements. After 2 min of milling the contact resistance is smaller than the sheet resistance of the pure aluminum film $R_{\square,Al} \sim 1.5\,\Omega/\square$. From these results we estimate a milling rate $\sim$ 1 nm/min to 2 nm/min of the native AlO_x layer by the argon milling process. Patterned features in the resist are widened by $\lesssim 1\,\mu m$, together with a roughening of the edges (see. Fig. 4.4). After 3 min of milling the bottom aluminum electrode is still continuous, indicating the robustness of the ion milling process to slight variations of the milling rate. However, by completely removing aluminum layers we estimate that the milling rate of pure aluminum is about a factor of ten bigger than that of AlO_x for the investigated cleaning process.

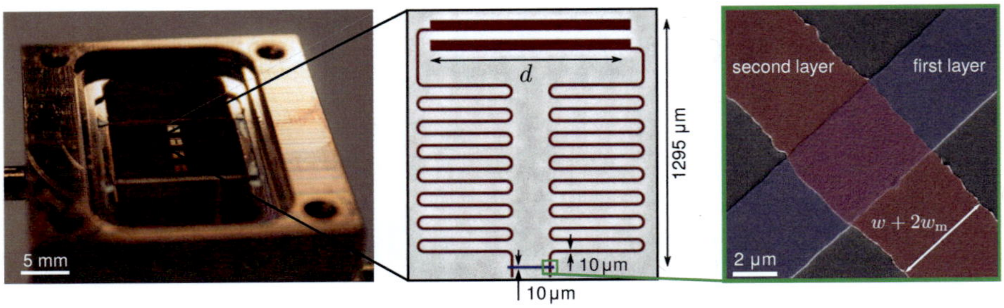

Figure 4.4: Resonator design for the test of argon ion milled overlap contacts. To decrease susceptibility to dielectric loss we employ the 3D rectangular waveguide sample holder (cf. Sec. 3.1.1). The zoom in shows a false colored optical image of a lumped element geometry resonator. The 12 mm long meandered inductor is identical between the four resonators on every chip, whereas we sweep the length d of the capacitor to design the resonant frequencies. False colored scanning electron image of the overlap contact area. As can be seen, the second metal layer (red), which is deposited following the argon milling process, is widened and has roughened edges.

In order to test the coherence of overlap contacts we perform a set of experiments on pure aluminum resonators in a 3D waveguide sample holder geometry. These resonators have two overlap contacts in a position of large current and small electric field, which we clean using the argon ion milling process. We compare the internal quality factors of resonators with milled contacts with those of resonators with nominally identical geometry, but fabricated in a single lithography step.

Figure 4.4 shows the resonator geometry, which consists of a $l = 12$ mm long meandered inductor with a width $w = 10\,\mu m$. Each set of samples consists of four resonators distributed in a frequency band of 300 MHz around $\sim$ 4.6 GHz. We achieve these different frequencies by varying the length d of the capacitor from 850 µm to 1000 µm in 50 µm steps. To decouple the resonators from the microwave environment, we intentionally design their resonant frequencies below the 3D waveguide sample holder cutoff frequency (cf. Sec. 3.1.1). Employing this strategy, we achieve coupling quality factors $Q_c \sim 10^6$.

To incorporation overlap contacts into the resonator we interrupt the meandered inductor in the middle and close the wire in a subsequent lithography step with in-situ argon ion milling immediately before metal deposition (cf. read meander and blue colored patch in Fig. 4.4). We characterize four fabrication variations, which are given in Table 4.3 in order to investigate different effects of the argon milling process. The first aluminum layer has a thickness of 30 nm, while the second has a thickness of 50 nm.

Figure 4.5 shows the measured resonant frequencies and internal quality factors of the four fabrication variations described in Tab. 4.3. Since the ion milling process increases the width of wires, which are deposited after the cleaning step, the resonant frequencies of sample A and C are lower compared to samples B and D. Through a wider wire the number of squares of the inductor l/w is reduced in samples B and D, which decreases the kinetic inductance and leads to higher frequencies. A widening of the strips $\sim 1\,\mu$m could explain the observed frequency shift on the order of 40 MHz, and is consistent with the widening found in SEM images.

Table 4.3: Fabrication sequence of experiments investigating the coherence of argon ion milled overlap contacts.

Sample	Fabrication step 1	Fabrication step 2
A	complete resonator	-
B	-	complete resonator
C	resonator, inductor interrupted	inductor closing bridge
D	inductor closing bridge	resonator, inductor interrupted

Since the kinetic inductance fraction in our pure aluminum resonators is on the order of 10 % to 15 %, we estimate that the change in kinetic inductance due to the different thicknesses of the aluminum layers should be on the order of 5 MHz.

The frequency difference between resonators fabricated in the same lithography step with and without overlap contacts is about 10 MHz (A and C) and 20 MHz (B and D). From the fact that resonant frequencies of samples with overlap contacts are higher than those of resonators without contacts we conclude a negligible kinetic inductance from the overlap contacts.

Based on the previous discussion, we would expect higher resonant frequencies of sample B compared to sample D since a larger part of the meander widens due to the milling in resonator D. Possible explanations for the observed higher frequencies of sample D could be fluctuations of the width of wires on the sample due to ion beam inhomogeneities or caused by non-uniform UV-exposure during the optical lithography process.

Figure 4.5 shows the mean internal quality factors of the resonators of each sample (solid line) and indicates the spread between the highest and lowest Q_{i} by the shaded area. The two single layer samples (A, B), and the sample, where the argon ion milling process is only applied to the area of the connecting patch (C) show internal quality factors $Q_\mathrm{i} > 10^6$ in the single photon regime. Sample D, where we apply the argon ion cleaning process also to the area of large electric field, shows a significantly lower internal quality factor on the order of 7×10^5. In this sample the milling might have caused a degradation of the substrate resulting in an increased dielectric loss similar to what has been reported in Ref. [156].

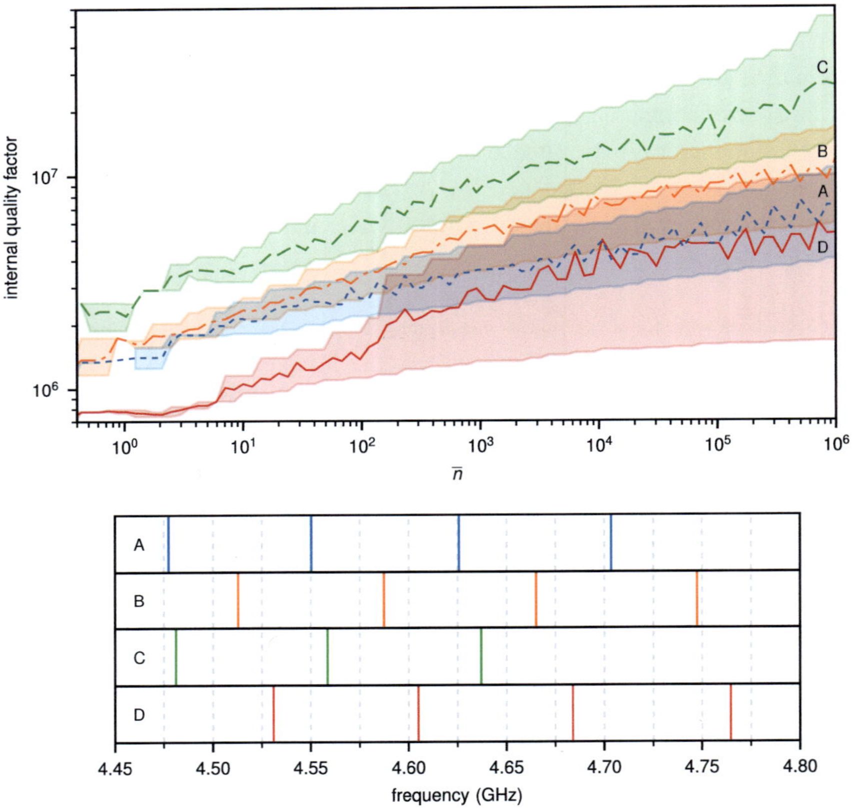

Figure 4.5: Internal quality factors as a function of the average number of circulating photons and resonant frequencies of the four samples. Samples which are fabricated in a single lithography step (A, B) as well as sample C, where only the overlap areas of the bottom layer are exposed to the argon ion milling process show internal quality factors larger than 10^6 in the single photon regime. Resonator D, where the substrate below the capacitor is exposed to the milling process, shows a lower internal quality factor. Resonators on sample C show higher resonant frequencies compared to sample A, indicating that the overlap junctions do not add significant kinetic inductance. The resonant frequencies of sample D show higher resonant frequencies compared to B, opposite to what is expected. We attribute this to fabrication inhomogeneities (for a more detailed discussion see the main text).

The mean internal quality factor of the samples with a bridge (B, D) allows to extract an upper bound on the residual resistance of the contacts by estimating the geometric inductance of a 1.6 cm long and 10 µm wide wire, $L \approx 33\,\mathrm{nH}$ and taking into account the kinetic inductance fraction $\alpha \sim 10\%$. Using the definition of the quality factor of a lossy inductor [37], we calculate a resistance $R = \omega L/Q_\mathrm{i} = 0.62\,\mathrm{m\Omega}$ for resonators with two overlap contacts, which have a mean internal quality factor $\overline{Q_\mathrm{i}} = 1.5 \times 10^6$ at a resonant frequency of 4.5 GHz. Taking into account the area of the overlap contacts we extract a residual resistivity limit for an overlap contact cleaned by an argon ion milling process of $R = (0.62\,\mathrm{m\Omega}/2)100\,\mathrm{\mu m^2} = 31\,\mathrm{m\Omega\,\mu m^2}$. We believe the presented argon ion milling process allows the fabrication of increasingly complex quantum circuit designs while maintaining a coherence on the level of $Q_\mathrm{i} \sim 10^6$.

4.2 Spectroscopy

In the following section we present results of standard two-tone spectroscopy measurement on two nominally identical grAl fluxonium qubits. Both samples are fabricated on the same chip with fluxonium qubits coupled to two readout resonators at different frequencies. For the fabrication we employed the three angle process as described in Sec. 4.1.3.

4.2.1 Sample 1

First, we find the readout resonator frequency at $f_\mathrm{r} = 7.278\,\mathrm{GHz}$, about 200 MHz lower than the designed value (cf. Tab. 4.1), which can be explained by a $\sim 5\%$ smaller kinetic inductance for a kinetic inductance fraction $\alpha \sim 1$. From a fit to the measured spectrum of the resonator we extract a total quality factor $Q_\mathrm{l} = 1.9 \times 10^3$, which is limited by a coupling quality factor $Q_\mathrm{c} = 2.3 \times 10^3$, as designed. Furthermore, from the fit we extract an internal quality factor $Q_\mathrm{i} = 11 \times 10^3$ for the readout resonator.

Figure 4.6 shows the result of a standard two-tone spectroscopy measurement. By monitoring the readout resonator response while sweeping a second continuous wave microwave tone in frequency we find the fluxonium qubit transition. Since the dispersive shift increases close to the resonator qubit anticrossing (cf. Fig. 1.9), we start spectroscopy close to this external magnetic flux point. A sweep of the bias current through the coil placed around the 3D rectangular waveguide sample holder while monitoring the resonant frequency of the readout resonator quickly reveals the bias currents at which resonance between the fluxonium and resonator occurs. This measurement is also already an indication for a functional fluxonium qubit. From measurements close to the anticrossing (see bottom right panel Fig. 4.6), we extrapolate the fluxonium qubit frequency at $\Phi_\mathrm{ext}/\Phi_0 = 0$ and $\Phi_\mathrm{ext}/\Phi_0 = 0.5$, which we subsequently measure to be 12.538 GHz, and 594 MHz, see bottom center and bottom left panel in Fig. 4.6 respectively. Using these measured points we perform a first preliminary fit of the fluxonium Hamiltonian to narrow down the search interval for the qubit frequency at the remaining external magnetic flux points. We measure in a frequency band ±50 MHz around the expected qubit frequency for external magnetic fluxes from $\sim -0.5\,\Phi_0$ to $0.5\,\Phi_0$ to minimize measurement time, while at the same time obtaining a fine resolution of the fluxonium spectrum.

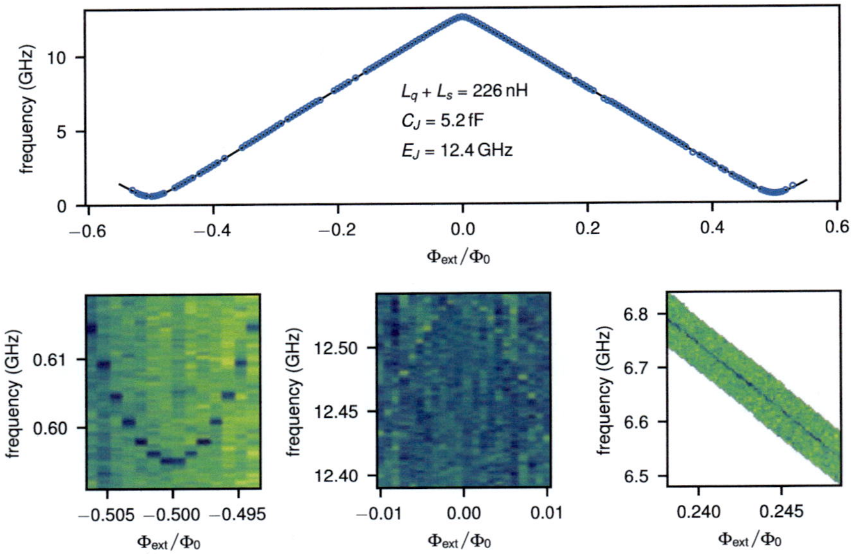

Figure 4.6: Spectrum of grAl fluxonium sample 1. The frequency of the first transition can be tuned from 595 MHz to 12.5 GHz as a function of the externally applied magnetic flux. From a numerical fit (solid line) to the measured data (blue points) we extract the qubit parameters, which are given in the plot. The bottom left color plot provides more details on one of the so called half-flux sweet spots, where the fluxonium frequency reaches a minimum and is to first order flux noise insensitive. The central bottom plot shows the zero flux sweet spot where the transition frequency reaches its maximum. Spectroscopy between the sweet spots over a wider frequency range does not show large anti-crossings which would be indicative of coupling to spurious modes or two-level systems [143].

Finally, we perform a numerical fit (black line) of the uncoupled fluxonium Hamiltonian to all measured points (blue cirlces), see Fig. 4.6. From this numerical fit we extract the fluxonium qubit parameters $L = 226\,\mathrm{nH}$, $C = 5.2\,\mathrm{fF}$, and $E_J/h = 12.4\,\mathrm{GHz}$. The fitted values for the Josephson energy and fluxonium qubit capacitance are in good agreement with the designed values (cf. Tab. 4.2). However, the fitted value of $L_q + L_s = 226\,\mathrm{nH}$ is $\sim 40\%$ smaller than designed, which is a significantly larger discrepancy compared to the inductance of the readout resonator. We attribute this difference to a variation in the geometry of the superinductor, possibly a larger width of the wire than designed. Further investigation of this difference would necessitate scanning electron microscopy images of the measured sample to obtain detailed information about the precise superinductor wire geometry.

In the entire fluxonium qubit spectrum, presented in Fig. 4.6 we did not observe any anti-crossings with spurious modes which would indicate strong coupling. However, aside from the half-flux and zero-flux sweet spot, and the qubit frequency interval 6.3 GHz to 6.8 GHz (see bottom panels Fig. 4.6), the resolution of the external magnetic flux was chosen such that it corresponds to a qubit frequency change of $\sim$ 140 MHz between two data points. Therefore, we can draw only limited conclusions from this observation.

The dispersive shift of the readout resonator was on the order of a few tens of kHz, about a factor of 5 to 10 smaller than designed. Consequently, the measurement signal is only on the order of $\sim$ 50 to 100 mrad, corresponding to an angle $\sim 5°$ between the ground and excited state points in the real-imaginary plane. This small separation mandates significant averaging of the readout resonator signal. Due to the small dispersive shift of the readout resonator, we were not able to identify any higher level transitions of the fluxonium qubit. However, from the good agreement between the measured lowest transition frequency and the numerical fit of the fluxonium Hamiltonian we conclude that the grAl superinductor indeed behaves as a linear, and lumped element inductor.

Figure 4.7 shows the spectrum of grAl fluxonium sample 1 following thermal cycling to room temperature twice and a total storage time in ambient atmosphere of six weeks between the cooldowns. By two-tone spectroscopy we measure the lowest fluxonium transition frequency at both sweet spots and close to the anti-crossing between resonator and qubit (blue points). A numerical fit to the measured data shows fluxonium circuit parameters in good agreement with the initially measured values. Therefore, we conclude negligible aging of the grAl superinductor and the Josephson junction (cf. Appendix A.2 for the employed process to remove resist residue, which could has been identified as a possible origin of aging [159]).

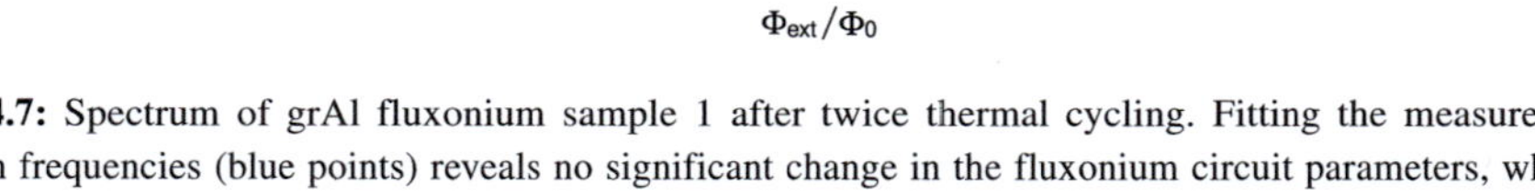

Figure 4.7: Spectrum of grAl fluxonium sample 1 after twice thermal cycling. Fitting the measured qubit transition frequencies (blue points) reveals no significant change in the fluxonium circuit parameters, which are given in the plot. From these results we conclude negligible aging of the grAl superinductor even if the sample is thermally cycled and stored in ambient atmosphere for a total time of six weeks.

4.2.2 Sample 2

The readout resonator of grAl fluxonium sample 2 has a resonant frequency $f_r = 7.053\,\text{GHz}$, and a total quality factor $Q_l = 1.4 \times 10^3$, similar to the values observed in the readout resonator of grAl fluxonium sample 1. We perform two-tone spectroscopy at the two extreme points of the fluxonium spectrum, and close to the resonator qubit anti-crossing, where the dispersive shift is on the order of 50 mrad. Blue points in Fig. 4.8 show the measured data and the black solid line is a numerical fit of the fluxonium Hamiltonian to the measured data. From the fit we extract the fluxonium parameters $L_s + L_q = 221\,\text{nH}$, $C = 5.8\,\text{fF}$, and $E_J/h = 13.2\,\text{GHz}$. The fitted parameters are comparable to those of grAl fluxonium sample 1, which is an identical fluxonium by design. At $\Phi_{ext}/\Phi_0 = 0$ (0.5) we measure a maximum (minimum) fluxonium qubit transition frequency of 13.093 GHz (406 MHz).

Due to the larger E_J in sample 2, the dispersive shift is even smaller than in sample 1 and on the order of a few kHz. Therefore, only a small number of measurement points are available and we are not able to measure the transition frequency to higher excited states. However, from the good agreement between numerical fit and measured data we again conclude that the grAl superinductor acts as modeled in the electrical schematic of the fluxonium circuit. Similar to sample 1, the inductance of the superinductor is about 40% smaller than expected, which could be due to a systematic difference between the designed and realized grAl wire geometry. To further investigate the origin of this difference scanning electron microscope images could provide further information about the wire dimensions.

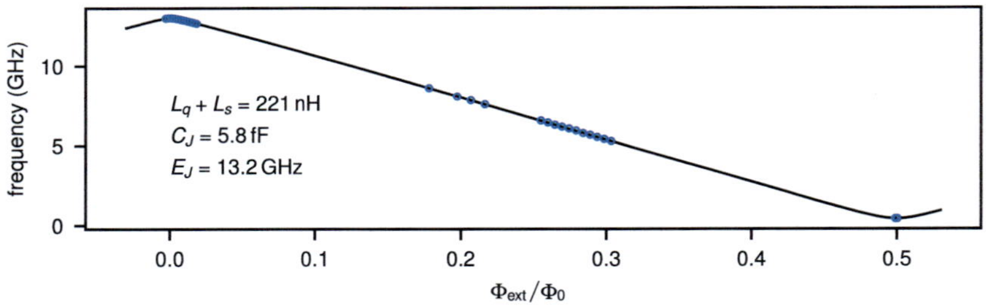

Figure 4.8: Spectroscopy of grAl fluxonium sample 2. Similarly to sample 1, the transition frequency can be tuned over $\sim 12\,\text{GHz}$ as a function of the external magnetic flux. The solid line is a numerical fit of the fluxonium Hamiltonian to the measured data (blue points). All fit parameters are given in the plot and are comparable to those of sample 1, which is by design an identical sample.

4.3 Coherence

In the following section we show the results of the coherence time measurements of grAl fluxonium samples 1 and 2. Due to the small dispersive shift of both samples, measurements were only possible at the two sweet spots. To conclude the section, we discuss possible limiting loss mechanisms for both samples based on the evaluation of T_1 relaxation channels as discussed in Sec. 1.2.3. We performed extensive measurements on sample 1, while sample 2 mostly serves as a control experiment and gives an indication about the reproducibility of the measured coherence times in sample 1.

4.3.1 Sample 1

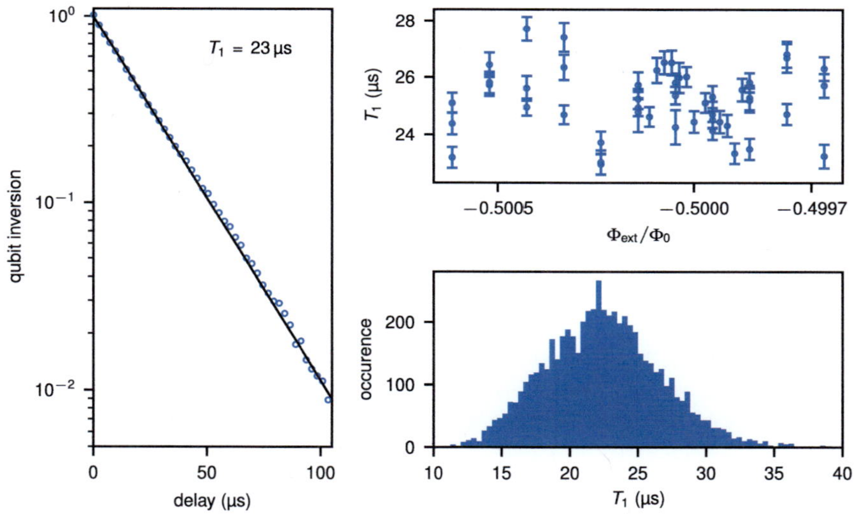

Figure 4.9: Energy relaxation time of fluxonium sample 1 at the half flux sweet spot. Exactly at half flux, we obtain $T_1 = 23\,\mu s$ for a single exponential decay. Measuring T_1 slightly detuned from the sweet spot does not show significantly lower energy relaxation times, indicating that T_1 is not limited by quasiparticle tunneling across the small Josephson junction. A repeated measurement over 17 h shows a mean $T_1 = 23\,\mu s$ with a standard deviation of $\pm 4\,\mu s$.

As the usual fluxonium qubit operation point is at an external flux of $\Phi_{ext}/\Phi_0 = 0.5$ we start the discussion of the grAl fluxonium coherence for this so called half-flux sweet spot. Figure 4.9 shows a T_1 measurement exactly at the half flux sweet spot averaged for $\sim 17\,h$. We calibrate the raw data of the T_1 relaxation measurement using the results of a Rabi oscillations experiment. This unambiguously determines the signal levels for the equilibrium qubit population and its inversion. Calibrating the raw T_1 data in this way, we observe a single exponential decay (blue points). The fit (black line) to the measured data yields $T_1 = 23\,\mu s$. The top right panel of Fig. 4.9 shows T_1 as a function of the external magnetic flux in a narrow interval. We extract energy relaxation times in the range of 22 µs to 28 µs. In this plot, error bars indicate the statistical uncertainty of the fit parameters. In the bottom right panel we show the histogram of ~ 6000 individual measurements, whose average is shown in the left panel. The shorter averaged T_1 are distributed around the mean of 23 µs with a standard deviation of $\pm 4\,\mu s$.

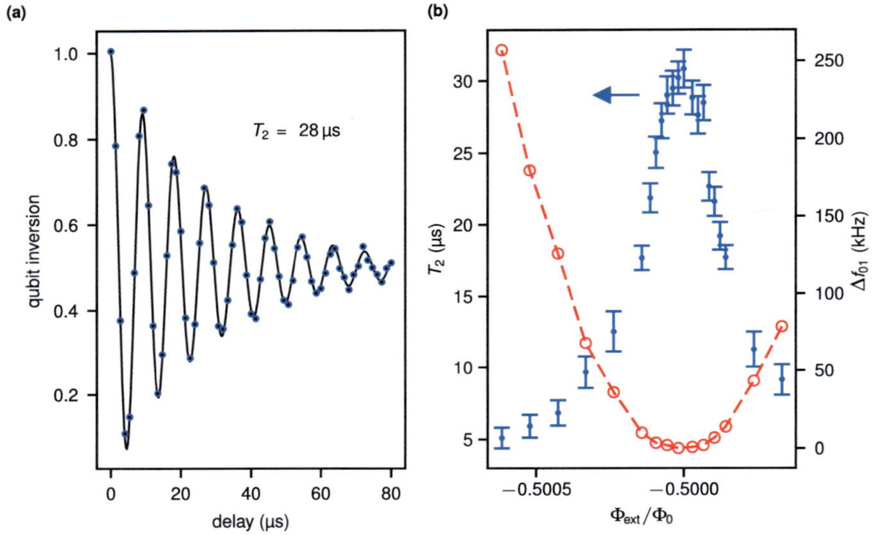

Figure 4.10: Coherence time T_2 of fluxonium sample 1 at the half flux sweet spot. From the fit of an exponentially damped cosine (solid line) to the measured data, we extract $T_2 = 28\,\mu s$ at $\Phi_{ext}/\Phi_0 = 0.5$. Due to the increasing susceptibility to flux noise we observe a decrease in T_2 by roughly a factor of 6 only $5\,m\Phi_0$ away from the sweet spot, which corresponds to a change in frequency of 250 kHz.

Figure 4.10 shows the coherence time T_2 measured by Ramsey fringes. The left panel shows a measurement at $\Phi_{ext}/\Phi_0 = 0.5$, where we drive the fluxonium qubit 110 kHz detuned from its transition frequency 594.37 MHz. By fitting an exponentially decaying cosine (black line) to the measured data (blue points) we extract a coherence time $T_2 = 28\,\mu s$. This coherence time is comparable to that of fluxonium qubits realized with Josephson junction superinductors [16, 17] or a NbTiN superinductor [26]. Since the fluxonium qubit spectrum is only exactly at the half-flux sweet spot first order flux noise insensitive, slightly moving away in flux already significantly decreases T_2. The right panel in Fig. 4.10 shows the decrease of T_2 by a factor ~ 6 only $5\,m\Phi_0$ away from the $\Phi_{ext}/\Phi_0 = 0.5$ sweet spot (blue points), which corresponds to a change in qubit frequency of 250 kHz (red circles). The error bars indicate the statistical uncertainty of the fitted value. From the strong flux dependence, we conclude that T_2 is limited by flux noise. Further away from the sweet spot, T_2 decreases to values on the order of 50 ns. The flux noise amplitude might be proportional to the length of the superinductor [160] and could be increased in our design compared to a Josephson junction fluxonium [161] due to the ~ 3 times longer loop. Furthermore, fluctuations of the kinetic inductance caused by two-level systems in disordered superconductors as recently proposed [162] could limit T_2 in our measurements.

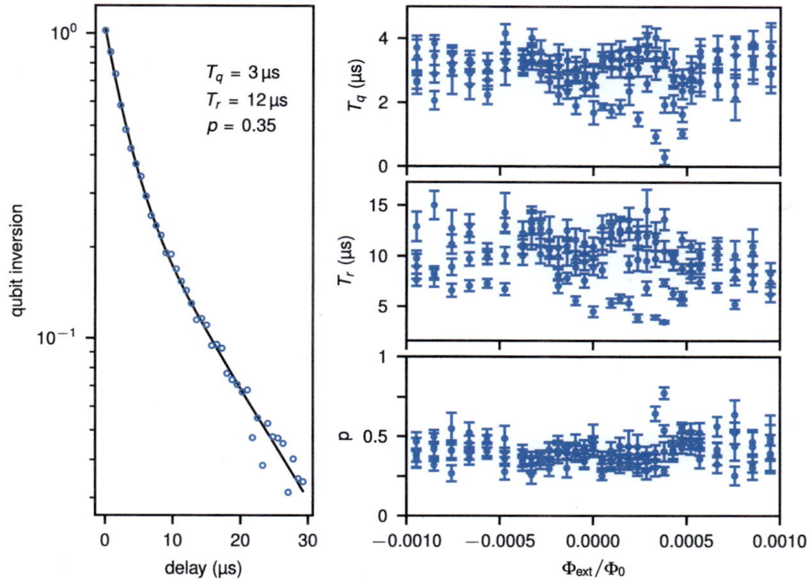

Figure 4.11: Energy relaxation time T_1 as a function of the external magnetic field close to the zero flux sweet spot. We observe a clear double exponential decay indicating a T_1 limited by quasiparticle tunneling across the Josephson junction [16, 104]. The quasiparticle relaxation time $T_q = 3\,\mu s$, while the residual decay time $T_r = 12\,\mu s$, which is observed in $p = 65\,\%$ of the measurements. Panels on the right side show the three fitting parameters as a function of the externally applied magnetic flux.

Figure 4.11 shows T_1 measured at $\Phi_{ext}/\Phi_0 = 0$. As for the measurements at the half-flux sweet spot, we record the data for T_1, T_2, and Rabi oscillations simultaneously by interleaving their respective measurement sequences. Using this approach, we are able to calibrate the data for the T_1 and T_2 measurements to full qubit population inversion. Employing the results of the Rabi oscillation measurement for the qubit state calibration, we observe a double exponential decay for the energy relaxation. This double exponential is a signature of T_1 limited by excess quasiparticles tunneling across the Josephson junction and has been reported previously [16, 43, 104]. We fit the double exponential decay with

$$P(t) = pe^{-\left(\frac{1}{T_q}+\frac{1}{T_r}\right)t} + (1-p)\,e^{-\frac{t}{T_r}}. \tag{4.1}$$

Using the formula we fit a relaxation process, which is limited by an additional relaxation mechanism with probability p. This loss channel, presumably due to the presence of excess quasiparticles in the Josephson junction vicinity, has a characteristic time constant T_q. In the remaining measurements, corresponding to a probability $1-p$ we observe a residual energy relaxation time T_r.

We fit Eq. 4.1 to the measured data at $\Phi_{\mathrm{ext}}/\Phi_0 = 0$ (see left panel of Fig. 4.11) and observe good agreement for fit parameters $T_q = 3\,\mu\mathrm{s}$, $T_r = 12\,\mu\mathrm{s}$, and $p = 0.35$. The right panel of Fig. 4.11 shows the three fit parameters as a function of the external magnetic flux. The error bars indicate the statistical uncertainty of the fit. We do not observe a dependence of the characteristic relaxation times or probability p on Φ_{ext}.

Figure 4.12 shows the coherence time T_2 in the vicinity of $\Phi_{\mathrm{ext}}/\Phi_0 = 0$. At the first order flux noise insensitive point (see left panel) we extract $T_2 = 4\,\mu\mathrm{s}$ from the fit of an exponentially decaying cosine (black line) to the measured data (blue points). Similarly to the half-flux sweet spot, away from $\Phi_{\mathrm{ext}}/\Phi_0 = 0$, T_2 decreases, while the qubit frequency only changes by 600 kHz relative to its maximum value of 12.538 GHz.

The significantly shorter coherence time $T_2 = 4\,\mu\mathrm{s}$ compared to 28 µs at $\Phi_{\mathrm{ext}}/\Phi_0 = 0.5$ could be due to the presence of excess quasiparticles, which cause qubit decoherence as explained in Refs. [34, 48].

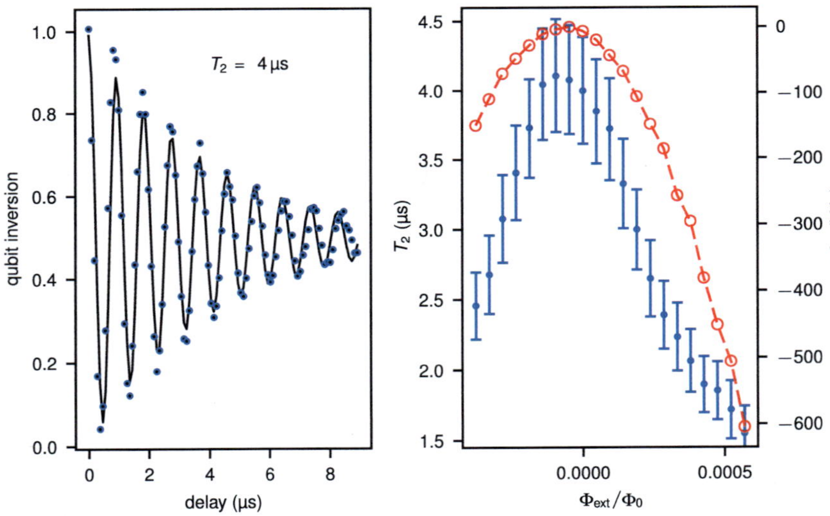

Figure 4.12: Coherence time T_2 of sample 1 at the zero flux sweet spot. We observe a maximum $T_2 = 4\,\mu\mathrm{s}$ exactly at zero flux and decreasing values away from the sweet spot, similarly to measurements at the half flux sweet spot, which is indicative of coherence limited by flux noise. The reduced T_2 at the sweet spot could be due to an increased excess quasiparticle density [48].

In Figure 4.13 we investigate capacitive, inductive, quasiparticle, and radiative loss as limiting loss mechanism for the energy relaxation time of grAl fluxonium sample 1. As already discussed and based on previously reported results [16, 43, 104], we attribute the shorter relaxation time $T_q = 3\,\mu s$ at $\Phi_{ext}/\Phi_0 = 0$ to quasiparticle tunneling across the Josephson junction, which we refer to as quasiparticle loss (red line). To explain the observed T_q a normalized excess quasiparticle density $x_{QP} = 1.5 \times 10^{-5}$ is necessary. The density is about one order of magnitude larger than previously observed in pure aluminum superconducting circuits and could be due to an effective quasiparticle trapping in the $Al/AlO_x/Al$ Josephson junction area, which is surrounded by grAl wires with $\sim 30\,\%$ higher superconducting gap. The residual $T_r = 12\,\mu s$ at $\Phi_{ext}/\Phi_0 = 0.5$ could be limited by dielectric loss in the fluxonium circuit, capping the dielectric quality factor at $Q_{cap} = 2.2 \times 10^5$. This loss could either occur due to the capacitor formed by the electrodes of the Josephson junction or due to other circuit areas which provide shunting capacitance.

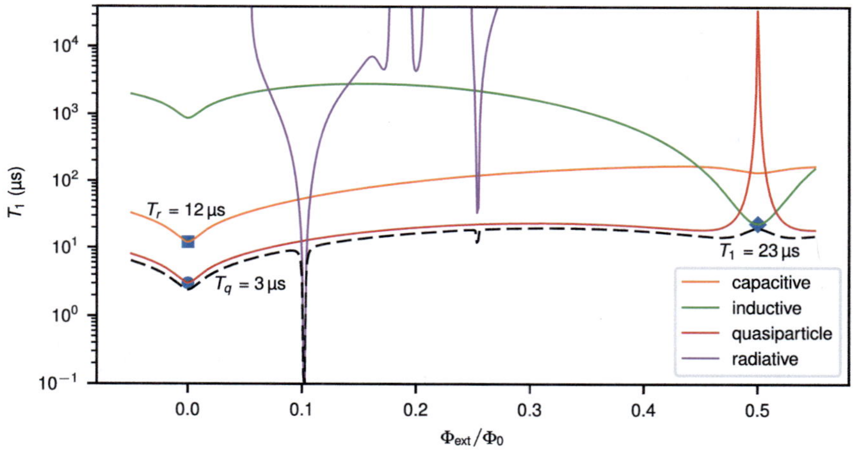

Figure 4.13: Loss channels as a function of the external magnetic flux. Following the methodology presented in Refs. [16, 29], we estimate the limiting loss mechanisms in grAl fluxonium sample 1 (cf. Tab. 1.2). We trace the limit due to capacitive loss (orange) assuming that the residual relaxation time $T_r = 12\,\mu s$ at $\Phi_{ext}/\Phi_0 = 0.0$ is dominated by capacitive loss. From this assumption we extract a quality factor of the capacitors $Q_{cap} = 2.2 \times 10^5$. Due to the double exponential decay at $\Phi_{ext}/\Phi_0 = 0.0$, we conclude that $T_q = 3\,\mu s$ is limited by excess quasiparticles tunneling across the small Josephson junction (red). Using the numerically calculated matrix elements (cf. Fig. 1.6), we estimate a normalized excess quasiparticle density $x_{QP} = 1.5 \times 10^{-5}$ at the Josephson junction. At $\Phi_{ext}/\Phi_0 = 0.5$, where dissipation due to quasiparticle tunneling is suppressed, we assume that an excess quasiparticle density $x_{QP} = 4.5 \times 10^{-7}$ in the superinductor limits $T_1 = 23\,\mu s$ (green). From a numerical simulation of the admittance towards the waveguide sample holder seen from the Josephson junction (cf. Sec. 1.2.3), we conclude that energy relaxation by emission into the electro-magnetic environment does not limit T_1 at the measured external magnetic field values (purple).

Finally, based on its functional dependence we believe inductive loss limits the fluxonium T_1 at $\Phi_{ext}/\Phi_0 = 0.5$ (green line). By inductive loss we refer to losses due to excess quasiparticles in the superinductor to distinguish them from losses which occur due to quasiparticle tunneling across the Josephson junction. An excess quasiparticle density $x_{QP} = 4.5 \times 10^{-7}$ in the superinductor, corresponding to an inductive quality factor $Q_{\text{ind}} = 1/x_{QP} = 2.2 \times 10^6$, could explain the measured T_1. This value is significantly higher compared to that in a Josephson junction superinductor [16], but also lower than observed in grAl strip resonators (cf. Sec. 3.4) and could be due to the slow relaxation time of quasiparticles in grAl [46]. The numerical simulation of the admittance seen from the Josephson junction into the 3D rectangular waveguide sample holder using finite elements methods permits to exclude radiative or Purcell loss [38] as a limiting loss mechanism for the measured energy relaxation times at the two fluxonium sweet spots (purple line). For a more detailed understanding of the limiting loss mechanisms in the grAl fluxonium more measurements of T_1, covering a much larger range of external magnetic fluxes are necessary, similar to the analysis performed in Ref. [16]. Due to the much smaller than designed dispersive shift of grAl fluxonium sample 1, a dedicated sample with larger readout signal would be needed.

4.3.2 Sample 2

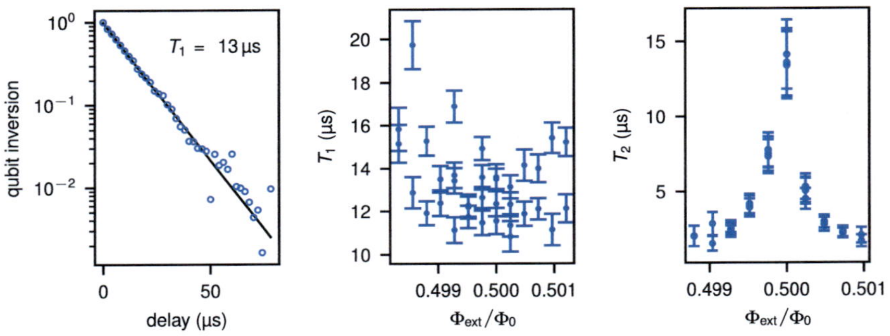

Figure 4.14: Coherence of grAl fluxonium sample 2 at $\Phi_{\text{ext}}/\Phi_0 = 0.5$. At the sweet spot we measure a single exponential energy decay $T_1 = 13\,\mu$s. As expected, T_1 does not show a dependence on the external magnetic flux. We attribute the observed fluctuations to varying T_1 in time, as recently also observed in transmon qubits [5–7]. In the right panel, we show T_2 as a function of the external magnetic field. At the sweet spot, we extract $T_2 = 13\,\mu$s from a Ramsey fringes measurement. Due to significantly increasing susceptibility to flux noise, we observe a rapidly decreasing T_2 away from the first order flux noise insensitive point at $\Phi_{\text{ext}}/\Phi_0 = 0.5$.

In this section we briefly present the measured coherence times of grAl fluxonium sample 2 at $\Phi_{\text{ext}}/\Phi_0 = 0.5$ and $\Phi_{\text{ext}}/\Phi_0 = 0$. Figure 4.14 shows T_1 and T_2 at the half-flux sweet spot. By interleaving the pulse sequences of T_1, T_2, and Rabi oscillation measurements, we simultaneously obtain the coherence times and are able to use the Rabi oscillation data to calibrate the qubit population inversion. At the half flux sweet spot, we observe a single exponential decay with $T_1 = 13\,\mu$s (see left panel in Fig. 4.14). The measured T_1 is independent of the external magnetic flux in the narrow interval shown in the central panel of Fig. 4.14 and we observe energy relaxation times between 10 µs and 20 µs. As already seen for sample 1 T_2 is strongly dependent on the external magnetic flux, due to the susceptibility to flux noise. At the first order flux noise insensitive point $\Phi_{ext}/\Phi_0 = 0.5$ T_2 reaches its maximum of 13 µs, comparable to the maximum T_2 measured in sample 1.

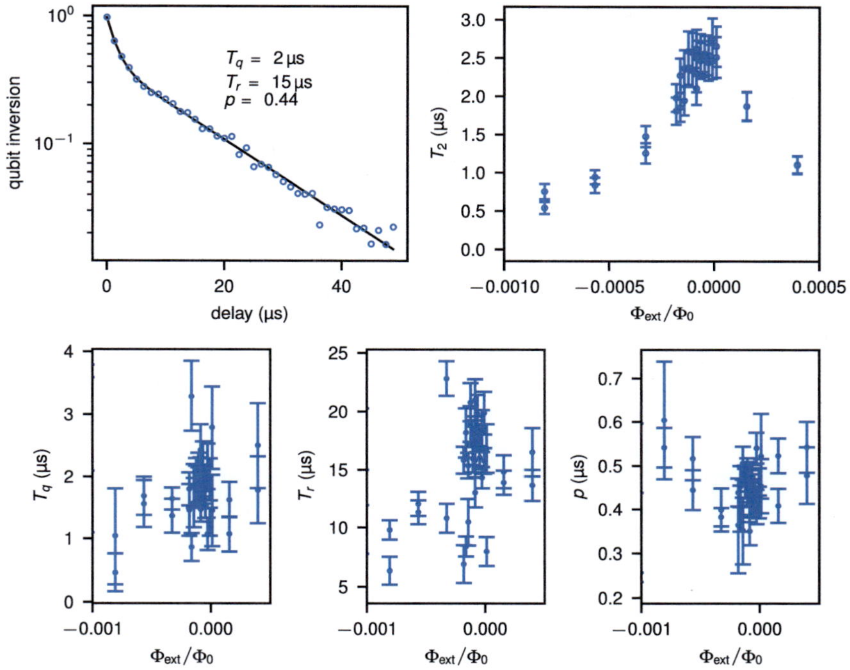

Figure 4.15: Coherence of grAl fluxonium sample 2 at $\Phi_{ext}/\Phi_0 = 0.0$. Similar to grAl fluxonium sample 1, we observe a double exponential decay for the energy relaxation time T_1 at the zero flux sweet spot, which is indicative of energy relaxation limited by the presence of excess quasiparticles at the Josephson junction. The top right panel shows T_2 as a function of the external magnetic flux, with the maximum value $T_2 = 2.5$ µs at the first order flux noise insensitive point $\Phi_{ext}/\Phi_0 = 0.0$. The bottom panels show T_q, T_r, and p as a function of the external magnetic flux.

Employing the same measurement approach as before, we quantify the coherence of grAl fluxonium sample 2 at the $\Phi_{ext}/\Phi_0 = 0$ sweet spot. Again, we observe a double exponential decay for T_1, which we fit using Eq. (4.1), as shown in the top left panel of Fig. 4.15. The fitted values $T_q = 2$ µs, $T_r = 15$ µs, and $p = 0.44$ are comparable to the parameters extracted from the energy relaxation measurements of sample 1. Similarly, T_2 is also strongly flux dependent and reaches a maximum $T_2 = 2.5$ µs, which decreases due to the increasing susceptibility to flux noise away from the sweet spot. The three lower panels of Fig. 4.15 show T_q, T_r, and p (left to right) as a function of the external magnetic flux.

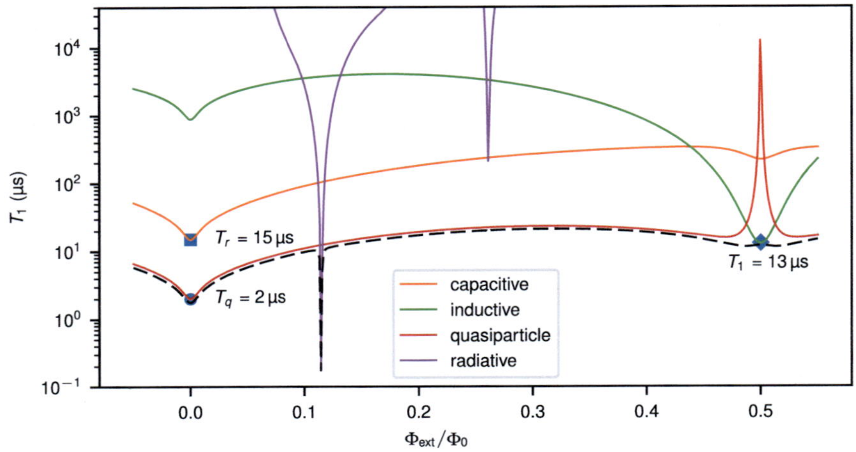

Figure 4.16: Loss channels as a function of the external magnetic flux. Following the methodology presented in Refs. [16, 29], we estimate the limiting loss mechanisms in grAl fluxonium sample 2 (cf. Tab. 1.2). We trace the limit due to capacitive loss (orange) assuming that the residual relaxation time $T_r = 15\,\mu s$ at $\Phi_{ext}/\Phi_0 = 0.0$ is dominated by capacitive loss. From this assumption we extract a quality factor of the capacitors $Q_{cap} = 2.2 \times 10^5$. Due to the double exponential decay at $\Phi_{ext}/\Phi_0 = 0.0$, we conclude that $T_q = 2\,\mu s$ is limited by excess quasiparticles tunneling across the small Josephson junction (red). Using the numerically calculated matrix elements (cf. Fig. 1.6), we estimate a normalized excess quasiparticle density $x_{QP} = 3.2 \times 10^{-5}$ at the Josephson junction. At $\Phi_{ext}/\Phi_0 = 0.5$, where dissipation due to quasiparticle tunneling is suppressed, we assume that an excess quasiparticle density $x_{QP} = 6.4 \times 10^{-7}$ in the superinductor limits $T_1 = 13\,\mu s$ (green). From a numerical simulation of the admittance towards the waveguide sample holder seen from the Josephson junction (cf. Sec. 1.2.3), we conclude that energy relaxation by emission into the electro magnetic environment does not limit T_1 at the measured external magnetic field values (purple).

Figure 4.16 provides an evaluation of the limiting loss mechanisms of the relaxation time of grAl fluxonium sample 2 based on the times measured at the two sweet spots. Due to the double exponential decay at Φ_{ext}/Φ_0, we again attribute T_q to the tunneling of excess quasiparticles across the Josephson junction, which leads to a normalized excess quasiparticle density $x_{QP} = 3.2 \times 10^{-5}$ (red line). Owing to the potential trapping of excess quasiparticles in the lower gap Al electrodes of the Al/AlO$_x$/Al Josephson junction the normalized excess quasiparticle density is about ten times higher compared to superconducting qubits fabricated from pure aluminum. The calculated x_{QP} is about a factor of two higher than previously reported for sample 1. The residual relaxation time T_r could be limited by dielectric loss in capacitive parts of the circuit, with a quality factor $Q_{cap} = 2.2 \times 10^5$ (orange line). At the $\Phi_{ext}/\Phi_0 = 0.5$ sweet spot $T_1 = 13\,\mu s$ is likely limited by dissipation due to quasiparticles in the grAl superinductor (green line). The estimated normalized excess quasiparticle density in the superinductor is $x_{QP} = 6.4 \times 10^{-7}$ and comparable to the density extracted for grAl fluxonium sample 1, which could be significantly increased compared to Josephson junction array superinductors [16] because of the slow relaxation time of quasiparticles in grAl thin films [46]. In agreement with sample 1 the observed energy relaxation times are not limited by radiative decay of the qubit into the electro magnetic environment provided by the 3D rectangular waveguide sample holder, as indicated by the purple line.

5 Conclusion & Outlook

In this thesis we have shown that granular aluminum is a promising material for superconducting quantum circuits with high characteristic impedance. Using various standard deposition techniques for pure aluminum in a slight oxygen atmosphere allows to fabricate granular aluminum films with normal state resistivities from $1\,\mu\Omega\,\mathrm{cm}$ to $10^4\,\mu\Omega\,\mathrm{cm}$. By tuning the resistivity of grAl films through the control of the oxygen partial pressure, the amenable kinetic inductance of grAl wires can be as high as nH/□. Thanks to that, and its straightforward deposition technique, which is fully compatible with the current standard Josephson junction fabrication by shadow evaporation, grAl wires reduce the complexity in design and fabrication required to realize superinductors. As we have shown, a grAl superinductor can also be in-situ integrated with a Al/AlOx/Al Josephson junction in a single lithography step. Based on these fabrication and design features superinductors might become a standard circuit element like capacitor, inductor, or Josephson junction in superconducting quantum information devices in the future.

In a first step we investigated grAl microwave resonators and extracted their internal quality factors to evaluate the dominating loss mechanisms at microwave frequencies in granular aluminum. Employing different geometries like standard coplanar waveguide resonators, lumped element kinetic inductance detectors, and rectangular strip resonators in a 3D waveguide sample holder, we tuned the energy participation ratio of the different interface regions and thereby the susceptibility of the resonators to dielectric loss. Our results show that grAl resonators with surface participation ratios $\gtrsim 10^{-3}$ are limited by dielectric loss in interface regions, with a loss tangent comparable to values reported in the literature for pure aluminum. In rectangular strip resonators in a 3D sample holder with further reduced surface participation ratio and internal quality factors on the order of 10^5, we attribute the limiting loss mechanism to excess quasiparticles. From a comparison with finite element methods simulations we extract a kinetic inductance of up-to $2\,\mathrm{nH}/\Box$ in these resonators, resulting in a kinetic inductance fraction $\alpha = L_{\mathrm{kinetic}}/L_{\mathrm{total}} \approx 1$, and consequently a participation ratio of unity for inductive loss. Based on these results we estimate a normalized excess quasiparticle density $x_{\mathrm{QP}} = 5 \times 10^{-6}$, which is in good agreement with previously reported normalized excess quasiparticle densities in pure aluminum quantum circuits. In grAl resonators with high kinetic inductance we observe stochastic quasiparticle bursts at an average rate of

$1/20\,\mathrm{s}$. Following such an event we observe a quasiparticle relaxation time on the order of seconds for resonators at a base temperature $T \sim 25\,\mathrm{mK}$, which is orders of magnitude longer than observed in kinetic inductance detectors fabricated from pure aluminum. The signatures of the observed bursts are reminiscent of high energy particle impacts in microwave kinetic inductance detectors. Despite these remarkably long quasiparticle relaxation times, the internal quality factors in grAl microwave resonators are comparable to superinductors realized using Josephson junction arrays.

To further investigate granular aluminum films as an option for the realization of superinductors in superconducting quantum circuits, we built a fluxonium qubit, i.e. a small Josephson junction shunted by a large superinductor. By adjusting the external magnetic flux threading the superinductor loop one can not only adjust the fluxonium qubit transition frequency but also its sensitivity to dissipation induced by dielectric loss, by excess quasiparticles either in the superinductor or at the small Josephson junction, or by radiative loss into strongly coupled electromagnetic modes. In two grAl fluxonium devices we measure coherence times T_1 up to $23\,\mu\mathrm{s}$ and T_2 up to $28\,\mu\mathrm{s}$, which are comparable to other current superconducting qubit implementations. As mentioned in the beginning of this Chapter, we believe that these results establish granular aluminum as a viable and versatile material for superconducting quantum circuits, which could allow the realization of increasingly complex superconducting quantum information processing devices.

Furthermore, our results indicate future directions of research to optimize superconducting circuits employing granular aluminum or other high kinetic inductance materials. As mentioned earlier, a normalized excess quasiparticle density on the order of 10^{-6} potentially limits the coherence of superconducting quantum circuits. In order to mitigate the influence of quasiparticles, possible approaches are quasiparticle or phonon traps to prevent them from reaching superconducting circuits, or the elucidation and subsequent reduction of quasiparticle producing events. Focusing on the sources of the quasiparticle bursts and systematically investigating their rate depending on shielding against infrared radiation, ambient radioactivity, or other particles capable of depositing energy $\gg 2\Delta_0$ could identify optimized strategies to decrease the normalized excess quasiparticle density. This would not only have implications for circuits employing high kinetic inductances, but also for integrated, multi-qubit superconducting quantum information devices. Based on the measured rate of quasiparticle bursts and the footprint of the resonators used to observe them, it is possible that these impacts happen in the substrate of the chip. Although there are already results in kinetic inductance detectors supporting this hypothesis, it would be interesting to perform such experiments in the hermetically shielded cryogenic setups used for superconducting qubits. If quasiparticle burst are indeed a result of high energy particle impacts in the substrate, qubit errors induced by hot non-equilibrium quasiparticles could be correlated in qubits fabricated on the same chip, which poses a challenge to current error correction protocols.

Possible mitigation strategies for the described impacts could be phonon traps of a lower gap material to prevent phonons with energy $> 2\Delta_0$ from reaching the qubits. Here, the enhanced superconducting gap of granular aluminum provides a possible test bed in combination with pure aluminum phonon traps to investigate the effectiveness of phonon traps depending on their location, difference in Δ_0 and other parameters. Finally granular aluminum could also be of interest for sensitive kinetic inductance detectors due to its high kinetic inductance and particularly slow quasiparticle relaxation times. These slow relaxation times are an interesting field of research themselves and future experiments employing resonators with different grAl film resistivities might prove useful in the understanding of the underlying physical process.

In conclusion granular aluminum is a versatile material for further developments of superconducting quantum information processing devices and sensitive detectors. Additionally, it might enable experiments to elucidate the sources of the observed large normalized excess quasiparticle density, or find possible mitigation strategies. Both aspects could become important factors for research towards quantum information processing, with granular aluminum potentially providing a basis for future quantum hardware developments.

Appendix

A Fabrication techniques

A.1 Granular aluminum deposition

To deposit grAl films we use a dedicated pure aluminum crucible in a commercial, controlled-angle *e*-beam evaporation machine (Plassys MEB 550S™), which we also use to fabricate shadow-angle evaporation Josephson junctions. The machine has two chambers, one that is used for oxidation and doubles as load-lock, and a second UHV chamber for metal evaporation. Pure aluminum is evaporated at a rate of 0.3 nm/s while we control the oxygen partial pressure in the oxidation chamber using a mass flow controller (MFC). Due to the continuous pumping on the chamber we obtain a dynamic oxygen pressure in the 10^{-5} mbar range, while in the evaporation chamber it is in the 10^{-6} mbar range. As shown in Fig. A.1, the grAl resistivity changes exponentially with the oxygen partial pressure (cf. Ref. [140]). The high sensitivity of the film resistivity to evaporation rate and oxygen partial pressure results in film resistivities reproducible within a factor of two from one evaporation to the next.

Granular aluminum deposition process flow

1. Pump the loadlock for 3 h down to a pressure in the range of 5×10^{-7} mbar,
2. Resist residue cleaning for 2 min (cf. Sec. A.2),
3. Titanium evaporation with closed sample shutter (0.2 nm/s for 2 min),
4. Regulate aluminum evaporation rate to 0.3 nm/s,
5. Regulate oxygen mass flow depending on the targeted resistivity (cf. Fig. A.1),
6. Open shutter,
7. Wait for desired grAl film thickness, close shutter, stop oxygen flow, ramp down aluminum evaporation rate,

8. Target rinsing (evaporate 20 nm of pure aluminum at 1 nm/s with the shutter closed and no oxygen).

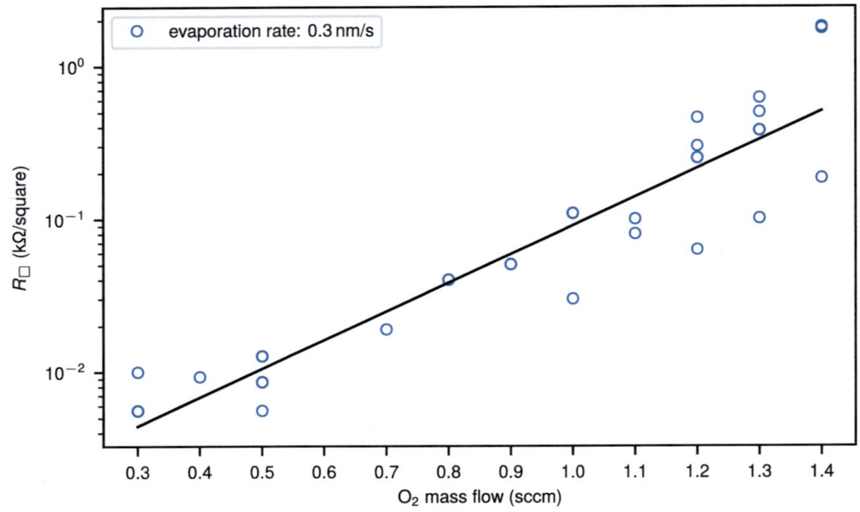

Figure A.1: Measured granular aluminum sheet resistance as a function of oxygen mass flow (blue circles). The functional dependence between sheet resistance $R_\square$ and oxygen mass flow during evaporation at 0.3 nm/s is well approximated by an exponential function $R_\square(x) = 0.001\,\mathrm{k}\Omega \times \exp\left[4.133x\,\mathrm{sccm}^{-1}\right]$ (black line). We attribute the observed fluctuations of $R_\square$ to variations of the evaporation rate smaller than the precision of the rate control, which is on the order of ± 0.03 nm/s.

Larger evaporation rates require a higher oxygen partial pressure to achieve the same film resistivities, which relaxes constraints on the resolution of the MFC to target a specific film resistivity. Furthermore, for optimal fabrication process reliability an in-situ measurement of the sheet resistance during film deposition would be a very valuable tool. Combined with control of the oxygen partial pressure during deposition this would allow a specific film resistance and thickness. An ability to stop film deposition at a targeted sheet resistance would also greatly facilitate device fabrication but compared to the previously mentioned approach would result in varying film thickness for the same sheet resistance and hence variations of the granular aluminum resistivity.

A.2 Resist residue cleaning

Previous results have shown that the increase of the tunnel barrier resistance of Josephson junctions fabricated by double-angle shadow evaporation techniques [59, 60] can be stopped at the price of a jump in resistance up to a factor of $\sim$4 by annealing the sample after fabrication [163]. Alternatively, this so-called *aging*, which happens during storage in ambient atmosphere, can be avoided by thoroughly removing any resist residue on the substrate in the Josephson junction area prior to deposition of the first electrode [159].

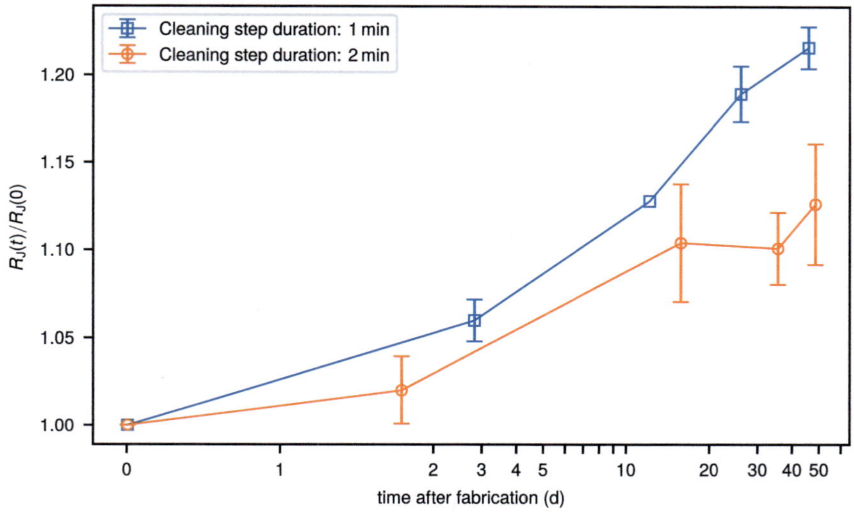

Figure A.2: Junction aging for resist residue cleaning times of 1 and 2 min. Applying the resist residue cleaning process for 2 min reduces the aging during sample storage in ambient atmosphere at room temperature to $\sim 10\%$. Error bars indicate two standard deviations around the mean of the two dataset, each consisting of approximately 10 individual measurements. Solid lines are a guide to the eye. For completely stable junction parameters the effectiveness of the cleaning step would have to be increased, either by extending the duration or by applying a plasma cleaning process with larger oxygen ratio similar to the one presented in Ref. [159].

For this purpose, we developed a cleaning process utilizing the Kaufman ion source of our controlled-angle evaporation tool, allowing in-situ substrate cleaning immediately before Josephson junction fabrication. The parameters of this process are as follows: 120 V acceleration voltage, 10 mA, 10 sccm O_2, 5 sccm Ar. This process results in an etch rate of ~0.35 nm/s for the top layer resist MicroChem™ A4 PMMA of our bilayer. Figure A.2 shows the results of two control experiments with different resist residue cleaning times. The process with previously mentioned parameters reduces aging to $\sim 10\%$ compared to junction aging of up to 300% for Josephson junctions without any cleaning. For all samples fabricated and measured during the course of this thesis the resist residue cleaning process was employed.

A.3 Electron-beam lithography lift-off

For a fast sample fabrication with only a small number of repetitions, we use an electron-beam lithography lift-off process, which employs a bilayer resist stack. Figure A.3 schematically illustrates the fabrication process. The bottom resist is a copolymer MMA (MicroChem™ EL-13), which we apply onto the used substrate using a spin coater. The resist is distributed over the entire substrate by spinning for 100 s at a speed of 2000 RPM. Afterwards, we perform a baking step of 5 min on a hot plate at 200 °C. This process results in a resist thickness $\sim$ 1 µm for the undercut layer. For the top resist we use a PMMA based resist (MicroChem™ A4), spun on top of the bottom layer at a rotation speed of 2000 RPM and using a spinning time of 5 min, which results in a resist thickness $\sim$ 300 nm. We use the same baking step of 5 min on a hot plate heated to 200 °C to harden the top resist. Finally, to avoid charging effects during the *e*-beam exposure step when using an insulating substrate, we sputter a gold layer with a thickness on the order of a few nm. This layer provides screening and avoids beam deflection by a charged resist or substrate.

For the patterning of our samples we use a 50 keV *e*-beam lithography system. Due to the moderate acceleration voltage, the electrons are scattered substantially in the bilayer resist stack, which leads to a significant beam broadening [164]. Additionally, backscattering of electrons in the substrate can occur. Since the bottom layer resist exhibits a sensitivity to *e*-beam exposure, which is roughly three times higher than the top layer, the two previously described processes result in a significant undercut $\sim$ 500 nm below the designed pattern in the top layer (cf. Fig. A.3). This process is later on deliberately used to pattern resist masks with asymmetric undercut [61] by adding areas of low *e*-beam exposure dose to allow the angled deposition of metal thin films (cf. Sec. 4.1.3).

After the sample was exposed to the *e*-beam write, the first step towards development is removing the gold layer using a *Lugol* solution, which is a mixture of potassium iodide and iodine with a ratio of KI:I $= 2:1$, solved in H_2O. Following the gold removal, we perform development of the exposed bilayer resist stack for 90 s in a $3:1$ mixture of IPE:H_2O, which is cooled down to 6 °C. This development technique increases the contrast of the PMMA and copolymer resists used in our fabrication process [165, 166].

The final step of the fabrication process after metal deposition is lift-off, performed in *N*-Methyl-2-pyrrolidone (NMP) or *N*-Ethyl-2-pyrrolidone (NEP) heated to 90 °C for $\sim$ 1 h, followed by a short, low power sonication step, and finished by consecutive rinsing steps in acetone, ethanol, and finally H_2O.

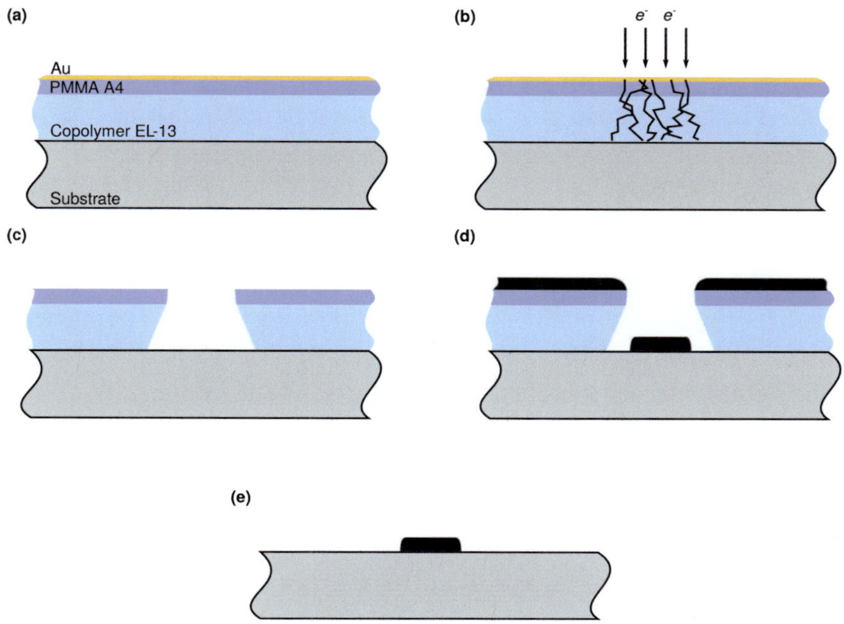

Figure A.3: Schematic process of e-beam lift-off lithography. **(a)** To allow the fabrication of controlled undercuts we employ a bilayer resist stack. A gold layer on top prevents charging on insulating substrates, which would deflect the electron writer and distort the pattern. **(b)** Exposure with an e-beam writer transfers the designed pattern into the bilayer resist stack. Depending on the acceleration voltage of the writer scattering of electrons is more or less severe and results in an unwanted undercut. **(c)** The exposed resist is developed for 90 s using a 3 : 1 mixture of IPA:H_2O, which is cooled down to 6 °C **(d)** Metal deposition on the developed resist stack transfers the design pattern into the superconducting thin film. **(e)** After lift-off in N-Methyl-2-pyrrolidone (NMP) or N-Ethyl-2-pyrrolidone (NEP) heated to 90 °C, all resist is removed. The pristine substrate with the designed pattern remains.

B Phenomenological model of quasiparticle activation

For completeness, in the following we present the derivation of the phenomenological model describing the quasiparticle (QP) dynamics in granular aluminum, as already give in the supplementary of Ref. [46].

In the presence of disorder, there are spatial variations of the superconducting order parameter. As discussed in Ref. [150], QPs can be localized at these spatial variations, which induce a small subgap tail in the electronic density of states. Here, we develop a phenomenological model similar to that introduced by Rothwarf and Taylor [151] that accounts for recombination of QPs, their generation, and transitions between localized and free (mobile) QPs.

With x_m and x_l we indicate the normalized density of mobile and localized QPs (x_m then corresponds to x_{qp}). We model the time evolution of these normalized densities by the following set of coupled, non-linear differential equations:

$$\dot{x}_m = -\Gamma_{mm}x_m^2 - \Gamma_{ml}x_m x_l - \Gamma_{\text{loc}}x_m + \Gamma_{\text{ex}}x_l + g_m, \quad (1)$$

$$\dot{x}_l = -\Gamma_{ll}x_l^2 - \Gamma_{ml}x_m x_l + \Gamma_{\text{loc}}x_m - \Gamma_{\text{ex}}x_l + g_l. \quad (2)$$

In both equations, the first two terms on the right hand side account for recombination between QPs of the same kind, with rates Γ_{mm} or Γ_{ll}, and between different types of QPs, with rate Γ_{ml}. The terms proportional to the rate Γ_{loc} describe the localization of mobile QPs, the rate Γ_{ex} the opposite excitation process. Finally, g_m and g_l are the generation rates of mobile and localized QPs. Note that assuming a constant density of localized QPs, with the identifications

$$\Gamma_{ml}x_l + \Gamma_{\text{loc}} \to s, \qquad \Gamma_{\text{ex}}x_l + g_m \to g, \qquad \Gamma_{mm} \to r, \quad (3)$$

Eq. (1) reduces to the phenomenological equation used for example in Ref. [43].

Since at long times after a QP burst, the deviations from the steady-state are small, we linearize the above equations by separating the small, time-dependent deviation from the steady state density:

$$x_m(t) = \bar{x}_m + \delta x_m(t), \qquad x_l(t) = \bar{x}_l + \delta x_l(t). \quad (4)$$

Keeping terms up to the first order we find

$$0 = -\Gamma_{mm}\bar{x}_m^2 - \Gamma_{ml}\bar{x}_m\bar{x}_l - \Gamma_{\text{loc}}\bar{x}_m + \Gamma_{\text{ex}}\bar{x}_l + g_m, \quad (5)$$

$$0 = -\Gamma_{ll}\bar{x}_l^2 - \Gamma_{ml}\bar{x}_m\bar{x}_l + \Gamma_{\text{loc}}\bar{x}_m - \Gamma_{\text{ex}}\bar{x}_l + g_l, \quad (6)$$

and

$$\begin{aligned}
\delta\dot{x}_m &= -2\Gamma_{mm}\bar{x}_m\delta x_m - \Gamma_{ml}\bar{x}_m\delta x_l - \Gamma_{ml}\bar{x}_l\delta x_m - \Gamma_{\text{loc}}\delta x_m + \Gamma_{\text{ex}}\delta x_l, &(7)\\
\delta\dot{x}_l &= -2\Gamma_{ll}\bar{x}_l\delta x_l - \Gamma_{ml}\bar{x}_m\delta x_l - \Gamma_{ml}\bar{x}_l\delta x_m + \Gamma_{\text{loc}}\delta x_m - \Gamma_{\text{ex}}\delta x_l. &(8)
\end{aligned}$$

The general model in Eqs. (1)-(2) can be simplified by considering the microscopic origin of the various rates. At low temperatures and small QP density, recombination and scattering rates are determined by electron-phonon interaction. For a zero-temperature phonon bath, the generation rates due to phonons would be exactly zero, but QPs can be also generated by photons or other elementary particles (e.g., protons [150]) of sufficient energy; assuming this energy to be large compared to the gap, we can set $g_l = 0$. Localization and recombination take place by phonon emission; based on Ref. [167], at low temperatures, we expect $\Gamma_{\text{loc}} > \Gamma_{mm}$, and since $x_m \ll 1$, we neglect the term proportional to Γ_{mm} (in fact, this approximation is applicable under the much less stringent condition $\Gamma_{mm}x_m \ll \Gamma_{\text{loc}}$). Later on, we will consider excitation due to photons in the resonator by setting

$$\Gamma_{\text{ex}} = \Gamma_0\bar{n}, \tag{9}$$

while neglecting photon emission by QPs in comparison to phonon emission. This approximation should be valid so long as $\Gamma_0\bar{n} \ll \Gamma_{\text{loc}}$.

B.1 Steady-state

Using the previously described simplifications, Eqs. (5)-(6) become

$$\begin{aligned}
0 &= -\Gamma_{ml}\bar{x}_m\bar{x}_l - \Gamma_{\text{loc}}\bar{x}_m + \Gamma_{\text{ex}}\bar{x}_l + g_m, &(10)\\
0 &= -\Gamma_{ll}\bar{x}_l^2 - \Gamma_{ml}\bar{x}_m\bar{x}_l + \Gamma_{\text{loc}}\bar{x}_m - \Gamma_{\text{ex}}\bar{x}_l. &(11)
\end{aligned}$$

Solving the last equation for $\bar{x}_l$ in terms of $\bar{x}_m$, we find

$$\bar{x}_l = \frac{\sqrt{(\Gamma_{ml}\bar{x}_m + \Gamma_{\text{ex}})^2 + 4\Gamma_{ll}\Gamma_{\text{loc}}\bar{x}_m} - (\Gamma_{ml}\bar{x}_m + \Gamma_{\text{ex}})}{2\Gamma_{ll}}. \tag{12}$$

With the further assumption

$$\Gamma_{ll}\Gamma_{\text{loc}} \ll \Gamma_{ml}^2\bar{x}_m, \tag{13}$$

we have

$$\bar{x}_l \simeq \frac{\Gamma_{\text{loc}}\bar{x}_m}{\Gamma_{ml}\bar{x}_m + \Gamma_{\text{ex}}}. \tag{14}$$

Substituting this expression into Eq. (10), we get

$$\bar{x}_m \simeq \frac{g_m}{2\Gamma_{\rm loc}}\frac{1}{2}\left(1+\sqrt{1+8\frac{\Gamma_{\rm loc}\Gamma_{\rm ex}}{g_m\Gamma_{ml}}}\right) = \frac{g_m}{2\Gamma_{\rm loc}}\left[1+\frac{1}{2}\left(\sqrt{1+4\gamma\bar{n}}-1\right)\right], \qquad \gamma = 2\frac{\Gamma_{\rm loc}\Gamma_0}{g_m\Gamma_{ml}}. \tag{15}$$

Then we can rewrite Eq. (14) as

$$\bar{x}_l = \frac{\Gamma_{\rm loc}}{\Gamma_{ml}}\frac{1}{1+\frac{\gamma\bar{n}}{1+\frac{1}{2}\left(\sqrt{1+4\gamma\bar{n}}-1\right)}}. \tag{16}$$

B.2 Decay rate

From Eqs. (7)-(8), we obtain a fast λ_+ and a slow one λ_-:

$$\lambda_\pm = \frac{1}{2}\left[a+b\pm\sqrt{(a-b)^2+4c}\right], \tag{17}$$

with

$$a = 2\Gamma_{ll}\bar{x}_l+\Gamma_{ml}\bar{x}_m+\Gamma_{\rm ex}, \qquad b = 2\Gamma_{mm}\bar{x}_m+\Gamma_{ml}\bar{x}_l+\Gamma_{\rm loc}, \qquad c = (\Gamma_{\rm loc}-\Gamma_{ml}\bar{x}_l)(\Gamma_{\rm ex}-\Gamma_{ml}\bar{x}_m). \tag{18}$$

A limiting regime is when $|c| \ll (a-b)^2$, in which case the two decay rates are approximately given by a and b. As we discuss in the next section, we expect in practice $a < b$, so a is the slow mode. Then under the assumption in Eq. (13), which leads to Eq. (14), we can neglect the first term in the definition of a with respect to the second term and we get

$$\lambda_- \simeq a \simeq \Gamma_{ml}\bar{x}_m+\Gamma_{\rm ex} \simeq \Gamma_0\left\{\frac{1}{\gamma}\left[1+\frac{1}{2}\left(\sqrt{1+4\gamma\bar{n}}-1\right)\right]+\bar{n}\right\}. \tag{19}$$

B.3 Comparing to experiments

The inverse quality factor can be generically written as the sum of the inverse of the quality factor at zero average photons Q_0 (due to both QPs present in the absence of photons and other loss mechanisms, such as dielectric losses) plus a photon number dependent part due to the change in the number of QPs. In principle, both localized and mobile QPs can contribute to the losses in a way proportional to their normalized density times coupling strength times final density of states. Since the change in quality factor is less than one order of magnitude, we expect the change in normalized density to be small, and to compare localized and mobile QPs we can use their zero-photon values. We also expect $\Gamma_{ml} < \Gamma_{mm}$, since the spatial overlap between localized and mobile QPs cannot be larger than that between mobile QPs. Then, the simplifying assumption $\Gamma_{mm}\bar{x}_m \ll \Gamma_{\mathrm{loc}}$ implies $\bar{x}_l > \bar{x}_m$. The final density of states is likely larger for localized QPs than for mobile QPs, for example if the former go from below the gap to just above the gap, while the latter always end at a higher energy above the gap, where the density of states is smaller. Therefore, if the coupling strengths to photons are similar, the localized QPs give a larger contribution to the losses (i.e., inverse quality factor), and using Eq. (16) we can write

$$\frac{1}{Q_i} = \frac{1}{Q_0} + \beta \left[\frac{1}{1 + \frac{\gamma \bar{n}}{1+\frac{1}{2}(\sqrt{1+4\gamma\bar{n}}-1)}} - 1 \right], \tag{20}$$

where β is proportional to $\Gamma_{\mathrm{loc}}/\Gamma_{ml}$ but also accounts for coupling strength and final density of states. Note that the formula cannot be extrapolated to very large $\bar{n}$ for two reasons: according to Eq. (15), the density of mobile QPs increases with $\bar{n}$; therefore their contribution to the quality factor can become relevant, and also neglecting the recombination between mobile QPs with respect to their localization will not hold anymore.

For the inverse of the decay time, we can proceed in a similar way and sum a photon-dependent part [from Eq. (19)] to a residual decay rate Γ_r, which includes both the zero-photon decay process which is part of the model, as well as other mechanisms not explicitly accounted for:

$$\frac{1}{\tau_{\mathrm{ss}}} = \Gamma_r + \Gamma_0 \left[\bar{n} + \frac{1}{2\gamma}\left(\sqrt{1+4\gamma\bar{n}} - 1\right)\right]. \tag{21}$$

Note that, since we have included the zero-photon contribution of Eq. (19) into Γ_r, we should always have $\Gamma_r \geq \Gamma_0/\gamma$.

Finally, we check for consistency of assumptions: let us use $\gamma \sim 1$ and $\Gamma_0 \sim 10^{-2}\,\mathrm{s}^{-1}$ (cf. Fig. 3.22). From Ref. [43] we have $\bar{x}_m \sim 10^{-6}$, $g_m \sim 10^{-4}\,\mathrm{s}^{-1}$ and $\Gamma_{mm} \sim 10^7\,\mathrm{s}^{-1}$. Using these values in Eq. (15) (with $\bar{n} = 0$) we estimate $\Gamma_{\mathrm{loc}} \sim 10^2\,\mathrm{s}^{-1}$, and using the definition of γ in the same equation we also estimate $\Gamma_{ml} \sim 10^4\,\mathrm{s}^{-1}$. Therefore, we find that indeed $\Gamma_{ml} < \Gamma_{mm}$, and also $\Gamma_{mm}\bar{x}_m \ll \Gamma_{\mathrm{loc}}$. Even for $\bar{n} \sim 10^2$ we have $\Gamma_0\bar{n} \ll \Gamma_{\mathrm{loc}}$. With these estimates, the assumption in Eq. (13) becomes $\Gamma_{ll} \ll 1\,\mathrm{s}^{-1}$, but to our knowledge there are no experimental data on this rate. Keeping this assumption, we can also estimate the quantities in Eq. (18): $a \sim \Gamma_0\,(1/\gamma + \bar{n}) \lesssim 1\,\mathrm{s}^{-1}$ (for $\bar{n} \lesssim 10^2$), $b > \Gamma_{\mathrm{loc}} \sim 10^2\,\mathrm{s}^{-1}$, and $|c| \lesssim \Gamma_0\Gamma_{\mathrm{loc}}\bar{n} \lesssim 10^2\,\mathrm{s}^{-2}$ (for $\bar{n} \lesssim 10^2$). These estimates verify the assumptions $a < b$ and $(a-b)^2 \gg |c|$.

C Experimental setup

C.1 Cryogenic microwave wiring

The use of superconducting materials and operating frequencies in the GHz regime mandate a temperature of the experimental setup T ßΔ_0/k_B, with the superconducting energy gap Δ_0 to suppress residual quasiparticle population of the superconductors. Furthermore $T < hf/k_\mathrm{B}$, with cicruit frequency f and Planck's constant h. For frequencies on the order of 5 GHz it follows $T < 240\,\mathrm{mK}$. Therefore, the measurements in this thesis were performed in commercial dilution cryostats with base temperatures $T_\mathrm{MXC} \approx 20\,\mathrm{mK}$. The main workhorse during the course of this thesis was an Oxford Instruments Kelvinox 100 commercial $^3\mathrm{He}/^4\mathrm{He}$ dilution cryostat. Since our lab is called the Black Forest Quantum lab, naturally we baptized the cryostat with the name of a tricky, and very smart bird: *Magpie*. Figure C.1 shows the inset of our cryostat.

Figure C.2 shows the initial wiring of *Magpie*. As can be seen in the schematics, the fridge provides two fully independent measurement lines allowing to perform two experiments simultaneously. Input signals on both lines are attenuated at different temperature stages of the cryostat to minimize Johnson-Nyquist noise and suppress residual photon population of the measured samples. Multiple commercial and home-made microwave filters ensure an attenuation $\lesssim -20\,\mathrm{dB}$ compared to pass band frequencies below the filter cutoff frequency of 8.2 GHz. After successful development and fabrication of parametric ultra-low noise dimer Josephson junction array amplifiers (DJJAA) [153], we adjusted the wiring of the cryostat to allow the operation of a DJJAA on one of the two measurement lines (see Fig. C.3). After interacting with the sample the signals pass an isolator used to shield the measured device from noise emitted back from the commercial low noise high electron mobility transistor amplifier (HEMT), which then subsequently amplifies signals by 43 dB.

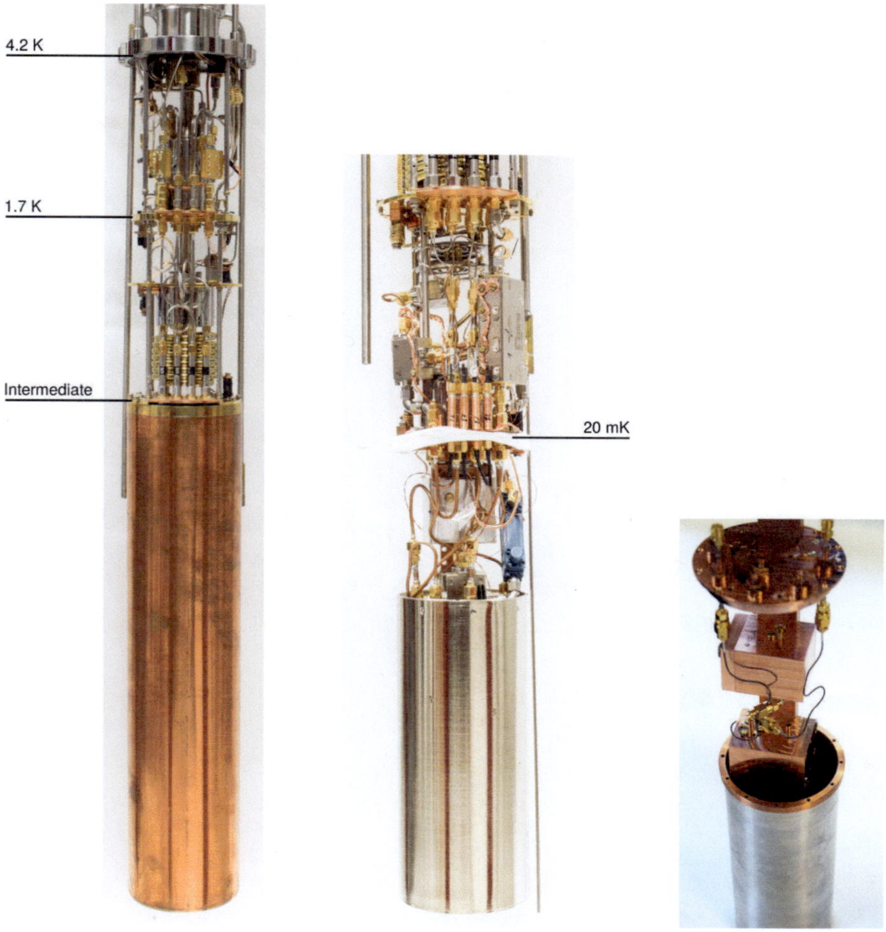

Figure C.1: Temperature stages and mounting setup in the Magpie fridge. The inset is enclosed in a stainless steel vacuum chamber (not shown), which is immersed in a liquid ^{4}He bath. Inside the vacuum chamber the dilution stages are mounted on three copper plates. All parts of the inset colder than the intermediate stage at around 600 mK are enclosed by a copper shield to decrease heat transport by radiation from the ^{4}He shield. Finally, samples are enclosed in successive μ-metal, superconducting, and copper shields to provide magnetic and residual radiation shielding.

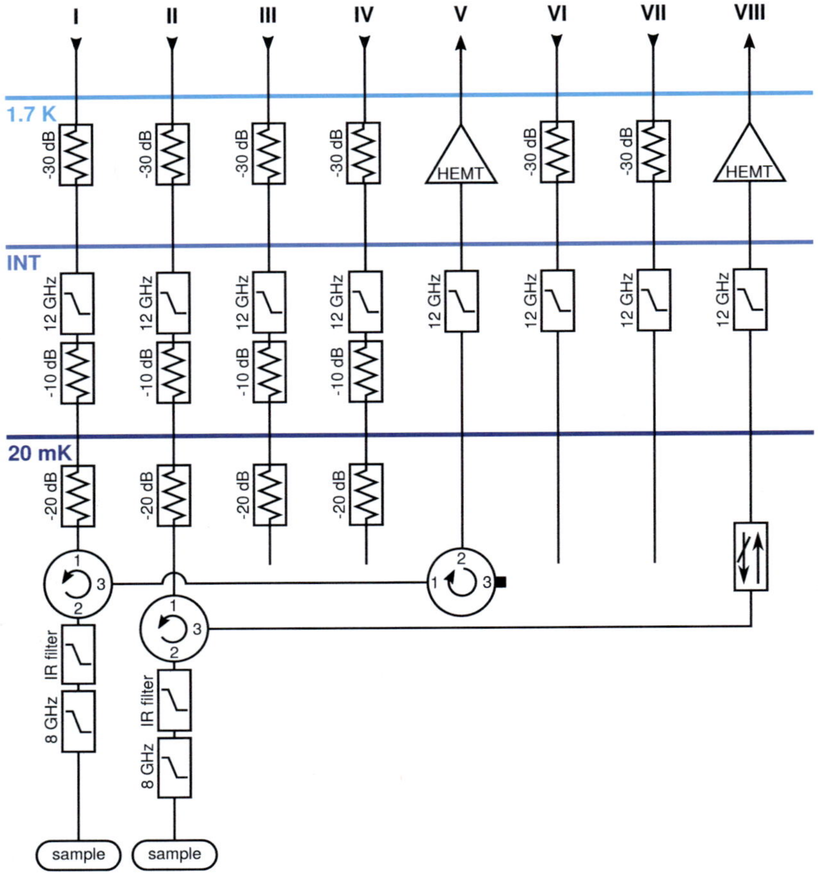

Figure C.2: Initial microwave wiring of the cryostat. Lines I and II are used to feed the readout microwave tone to the samples and are heavily attenuated and filtered by commercial (8 GHz and 12 GHz) low-pass and home made infrared (IR) filters. Lines III and IV provide a separate path for microwave tones used to manipulate the quantum state of the measured samples. Both output lines (V and VIII) are equipped with isolators to provide at least 20 dB of attenuation for signals traveling back from the commercial high electron mobility transistor amplifiers (HEMT).

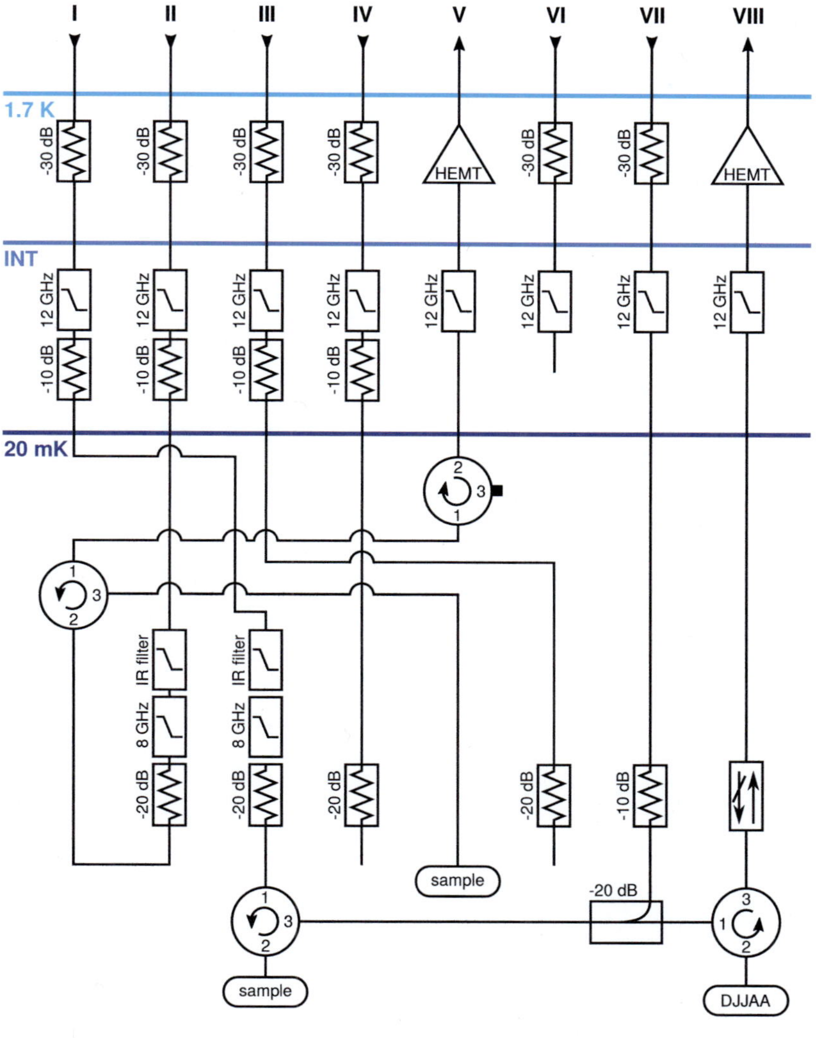

Figure C.3: Adjusted microwave wiring of the cryostat to enable the use of a parametric dimer Josephson junction array amplifier (DJJAA) [153]. Lines I and II are used to feed the readout microwave tone to the samples and are heavily attenuated and filtered by commercial (8 GHz and 12 GHz) low-pass and home made infrared (IR) filters. Lines III and IV provide a separate path for microwave tones used to manipulate the quantum state of the measured samples, while Line VII has a lower attenuation and is used to apply the pump tone for the near quantum limited DJJAA. Both output lines (V and VIII) are equipped with isolators to provide at least 20 dB of attenuation for signals traveling back from the commercial high electron mobility transistor amplifiers (HEMT).

C.2 Timedomain setup

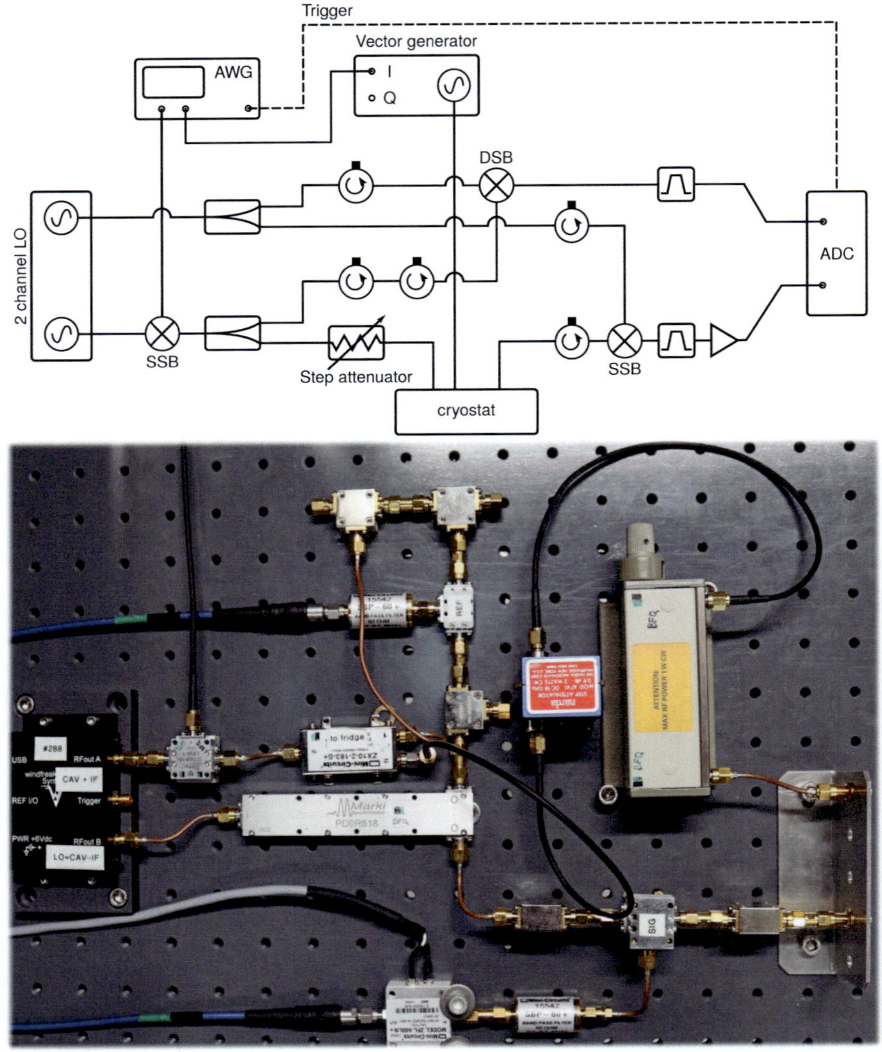

Figure C.4: Image and schamtic of the time domain setup built during the course of the thesis. We employ an interferometric setup to be insensitive of phase fluctuations occurring before the common reference plane, which is defined by the two splitters. One half of the signal travels through the cryogenic measurement setup and interacts with the sample, while the reference part of the signal only passes room temperature components. Signal and reference pulse are digitized by a 1.25 Gs/s analog digital converter (ADC). The reference pulse is used to digitally demodulate the signal pulse into its two quadratures, which constitute the readout signal. Multiple isolators in the room temperature setup suppress unintentional interference between both signal arms. Single sideband (SSB) microwave mixers are used for convenience and to suppress noise compared to double sideband mixers (DSB).

Bibliography

[1] R. P. Feynman: *Simulating physics with computers*, International Journal of Theoretical Physics **21**, 467–488 (1982), DOI: 10.1007/BF02650179.

[2] M. H. Devoret and R. J. Schoelkopf: *Superconducting circuits for quantum information: An outlook*, Science **339**, 1169–1174 (2013), DOI: 10.1126/science.1231930.

[3] Y. Nakamura, Y. A. Pashkin, and J. S. Tsai: *Coherent control of macroscopic quantum states in a single-cooper-pair box*, Nature **398**, 786–788 (1999), DOI: 10.1038/19718.

[4] R. Barends, J. Kelly, *et al.*: *Superconducting quantum circuits at the surface code threshold for fault tolerance*, Nature **508**, 500–503 (2014), DOI: 10.1038/nature13171.

[5] P. V. Klimov, J. Kelly, *et al.*: *Fluctuations of energy-relaxation times in superconducting qubits*, Physical Review Letters **121**, 090502 (2018), DOI: 10.1103/PhysRevLett.121.090502.

[6] J. Burnett, A. Bengtsson, M. Scigliuzzo, D. Niepce, M. Kudra, P. Delsing, and J. Bylander: *Decoherence benchmarking of superconducting qubits*, arXiv:1901.04417 [cond-mat.supr-con], (2019), [Online]. Available: https://arxiv.org/abs/1901.04417.

[7] S. Schlör, J. Lisenfeld, C. Müller, A. Schneider, D. P. Pappas, A. V. Ustinov, and M. Weides: *Correlating decoherence in transmon qubits: Low frequency noise by single fluctuators*, arXiv:1901.05352 [cond-mat.supr-con], (2019), [Online]. Available: https://arxiv.org/abs/1901.05352.

[8] J. Koch, T. M. Yu, J. Gambetta, A. A. Houck, D. I. Schuster, J. Majer, A. Blais, M. H. Devoret, S. M. Girvin, and R. J. Schoelkopf: *Charge-insensitive qubit design derived from the Cooper pair box*, Phys. Rev. A **76**, 42319 (2007), DOI: 10.1103/PhysRevA.76.042319.

[9] A. G. Fowler, M. Mariantoni, J. M. Martinis, and A. N. Cleland: *Surface codes: Towards practical large-scale quantum computation*, Physical Review A **86**, 032324 (2012), DOI: 10.1103/PhysRevA.86.032324.

[10] E. Campbell, A. Khurana, and A. Montanaro: *Applying quantum algorithms to constraint satisfaction problems*, arXiv:1810.05582 [quant-ph], (2018), [Online]. Available: `https://arxiv.org/abs/1810.05582`.

[11] Z. Leghtas, G. Kirchmair, B. Vlastakis, R. J. Schoelkopf, M. H. Devoret, and M. Mirrahimi: *Hardware-efficient autonomous quantum memory protection*, Physical Review Letters **111**, 120501 (2013), DOI: `10.1103/PhysRevLett.111.120501`.

[12] B. Vlastakis, G. Kirchmair, Z. Leghtas, S. E. Nigg, L. Frunzio, S. M. Girvin, M. Mirrahimi, M. H. Devoret, and R. J. Schoelkopf: *Deterministically encoding quantum information using 100-photon schrodinger cat states*, Science **342**, 607–610 (2013), DOI: `10.1126/science.1243289`.

[13] M. Mirrahimi, Z. Leghtas, V. V. Albert, S. Touzard, R. J. Schoelkopf, L. Jiang, and M. H. Devoret: *Dynamically protected cat-qubits: a new paradigm for universal quantum computation*, New Journal of Physics **16**, 045014 (2014), DOI: `10.1088/1367-2630/16/4/045014`.

[14] A. Kitaev: *Protected qubit based on a superconducting current mirror*, arXiv:cond-mat/0609441 [cond-mat.mes-hall], (2006), [Online]. Available: `https://arxiv.org/abs/cond-mat/0609441`.

[15] V. E. Manucharyan, J. Koch, L. I. Glazman, and M. H. Devoret: *Fluxonium: Single cooper-pair circuit free of charge offsets*, Science **326**, 113–116 (2009), DOI: `10.1126/science.1175552`.

[16] I. M. Pop, K. Geerlings, G. Catelani, R. J. Schoelkopf, L. I. Glazman, and M. H. Devoret: *Coherent Suppression of Electromagnetic Dissipation Due to Superconducting Quasiparticles*, Nature (London) **508**, 369–372 (2014), DOI: `10.1038/nature13017`.

[17] N. Earnest, S. Chakram, Y. Lu, N. Irons, R. K. Naik, N. Leung, L. Ocola, D. A. Czaplewski, B. Baker, J. Lawrence, J. Koch, and D. I. Schuster: *Realization of a {λ} system with metastable states of a capacitively shunted fluxonium*, Phys. Rev. Lett. **120**, (2018), DOI: `10.1103/PhysRevLett.120.150504`.

[18] L. B. Nguyen, Y.-H. Lin, A. Somoroff, R. Mencia, N. Grabon, and V. E. Manucharyan: *The high-coherence fluxonium qubit*, arXiv:1810.11006 [quant-ph], (2018), [Online]. Available: `https://arxiv.org/abs/1810.11006`.

[19] B. Josephson: *Possible new effects in superconductive tunnelling*, Physics Letters **1**, 251–253 (1962), DOI: `10.1016/0031-9163(62)91369-0`.

[20] A. Shnirman, G. Schön, and Z. Hermon: *Quantum manipulations of small josephson junctions*, Physical Review Letters **79**, 2371–2374 (1997), DOI: `10.1103/PhysRevLett.79.2371`.

[21] M. H. Devoret, "Quantum fluctuations in electrical circuits", *Quantum Fluctuations*, ser. Les Houches, Elsevier Science, 1995, pp. 351–386.

[22] E. Lucero, R. Barends, Y. Chen, J. Kelly, M. Mariantoni, A. Megrant, P. O'Malley, D. Sank, A. Vainsencher, J. Wenner, T. White, Y. Yin, A. N. Cleland, and J. M. Martinis: *Computing prime factors with a Josephson phase qubit quantum processor*, Nature Physics **8**, 719–723 (2012), DOI: 10.1038/nphys2385.

[23] J. Kelly, R. Barends, *et al.*: *State preservation by repetitive error detection in a superconducting quantum circuit*, Nature **519**, 66–69 (2015), DOI: 10.1038/nature14270.

[24] N. K. Langford, R. Sagastizabal, M. Kounalakis, C. Dickel, A. Bruno, F. Luthi, D. J. Thoen, A. Endo, and L. DiCarlo: *Experimentally simulating the dynamics of quantum light and matter at deep-strong coupling*, Nature Communications **8**, 1715 (2017), DOI: 10.1038/s41467-017-01061-x.

[25] S. M. Girvin, "Circuit QED: superconducting qubits coupled to microwave photons", *Quantum Machines: Measurement and Control of Engineered Quantum Systems*, M. Devoret, B. Huard, R. Schoelkopf, and L. F. Cugliandolo, Eds., Oxford University Press, 2014, pp. 113–256, DOI: 10.1093/acprof:oso/9780199681181.003.0003.

[26] T. M. Hazard, A. Gyenis, A. Di Paolo, A. T. Asfaw, S. A. Lyon, A. Blais, and A. A. Houck: *Nanowire superinductance fluxonium qubit*, Physical Review Letters **122**, (2019), DOI: 10.1103/PhysRevLett.122.010504.

[27] J. Koch, V. E. Manucharyan, M. H. Devoret, and L. I. Glazman: *Charging effects in the inductively shunted Josephson junction*, arXiv:0902.2980 [cond-mat.mes-hall], (2009), [Online]. Available: http://arxiv.org/abs/0902.2980.

[28] C. Cohen-Tannoudji, B. Diu, F. Laloë, J. Streubel, and J. Balla, *Quantenmechanik*, 2., durchges. und verb. Aufl. Berlin New York: de Gruyter, 1999, 2 pp.

[29] N. A. Masluk, *Reducing the losses of the fluxonium artificial atom*, Dissertation, Yale University, New Haven, CT, 2012.

[30] W. C. Smith, A. Kou, U. Vool, I. M. Pop, L. Frunzio, R. J. Schoelkopf, and M. H. Devoret: *Fluxonium-resonator system in the nonperturbative regime*, Phys. Rev. B **94**, 144507 (2016), DOI: https://doi.org/10.1103/PhysRevB.94.144507.

[31] D. Zwillinger, *Table of Integrals, Series, and Products.* Burlington: Elsevier, 2007, [Online]. Available: http://public.eblib.com/choice/publicfullrecord.aspx?p=287927 (visited on 04/08/2019).

[32] H. Paik, D. I. Schuster, L. S. Bishop, G. Kirchmair, G. Catelani, A. P. Sears, B. R. Johnson, M. J. Reagor, L. Frunzio, L. I. Glazman, S. M. Girvin, M. H. Devoret, and R. J. Schoelkopf: *Observation of high coherence in josephson junction qubits measured in a three-dimensional circuit QED architecture*, Physical Review Letters **107**, (2011), DOI: 10.1103/PhysRevLett.107.240501.

[33] R. J. Schoelkopf, A. A. Clerk, S. M. Girvin, K. W. Lehnert, and M. H. Devoret, "Qubits as Spectrometers of Quantum Noise", *Quantum Noise in Mesoscopic Physics*, Y. V. Nazarov, Ed., Dordrecht: Springer Netherlands, 2003, pp. 175–203, DOI: 10.1007/978-94-010-0089-5_9.

[34] G. Catelani, R. J. Schoelkopf, M. H. Devoret, and L. I. Glazman: *Relaxation and frequency shifts induced by quasiparticles in superconducting qubits*, Physical Review B **84**, (2011), DOI: 10.1103/PhysRevB.84.064517.

[35] G. Catelani, J. Koch, L. Frunzio, R. J. Schoelkopf, M. H. Devoret, and L. I. Glazman: *Quasiparticle relaxation of superconducting qubits in the presence of flux*, Physical Review Letters **106**, (2011), DOI: 10.1103/PhysRevLett.106.077002.

[36] N. A. Masluk, I. M. Pop, A. Kamal, Z. K. Minev, and M. H. Devoret: *Microwave characterization of josephson junction arrays: Implementing a low loss superinductance*, Physical Review Letters **109**, (2012), DOI: 10.1103/PhysRevLett.109.137002.

[37] D. M. Pozar, *Microwave engineering*, 4th ed. Hoboken, NJ: Wiley, 2012, 732 pp.

[38] E. M. Purcell: *Spontaneous emission probabilities at radio frequencies*, Physical Review **69**, 674–674 (1946), DOI: 10.1103/PhysRev.69.674.

[39] J. Aumentado, M. W. Keller, J. M. Martinis, and M. H. Devoret: *Nonequilibrium quasiparticles and 2 e periodicity in single-cooper-pair transistors*, Physical Review Letters **92**, (2004), DOI: 10.1103/PhysRevLett.92.066802.

[40] P. J. de Visser, J. J. A. Baselmans, P. Diener, S. J. C. Yates, A. Endo, and T. M. Klapwijk: *Number fluctuations of sparse quasiparticles in a superconductor*, Physical Review Letters **106**, (2011), DOI: 10.1103/PhysRevLett.106.167004.

[41] V. F. Maisi, S. V. Lotkhov, A. Kemppinen, A. Heimes, J. T. Muhonen, and J. P. Pekola: *Excitation of single quasiparticles in a small superconducting al island connected to normal-metal leads by tunnel junctions*, Physical Review Letters **111**, (2013), DOI: 10.1103/PhysRevLett.111.147001.

[42] E. M. Levenson-Falk, F. Kos, R. Vijay, L. Glazman, and I. Siddiqi: *Single-quasiparticle trapping in aluminum nanobridge josephson junctions*, Physical Review Letters **112**, (2014), DOI: 10.1103/PhysRevLett.112.047002.

[43] C. Wang, Y. Y. Gao, I. M. Pop, U. Vool, C. Axline, T. Brecht, R. W. Heeres, L. Frunzio, M. H. Devoret, G. Catelani, L. I. Glazman, and R. J. Schoelkopf: *Measurement and control of quasiparticle dynamics in a superconducting qubit*, Nature Communications **5**, 5836 (2014), DOI: 10.1038/ncomms6836.

[44] A. Bilmes, S. Zanker, A. Heimes, M. Marthaler, G. Schön, G. Weiss, A. V. Ustinov, and J. Lisenfeld: *Electronic decoherence of two-level systems in a josephson junction*, Physical Review B **96**, (2017), DOI: 10.1103/PhysRevB.96.064504.

[45] K. Serniak, M. Hays, G. de Lange, S. Diamond, S. Shankar, L. D. Burkhart, L. Frunzio, M. Houzet, and M. H. Devoret: *Hot nonequilibrium quasiparticles in transmon qubits*, Physical Review Letters **121**, 157701 (2018), DOI: 10.1103/PhysRevLett.121.157701.

[46] L. Grünhaupt, N. Maleeva, S. T. Skacel, M. Calvo, F. Levy-Betrand, A. V. Ustinov, H. Rotzinger, A. Monfardini, G. Catelani, and I. M. Pop: *Loss Mechanisms and Quasiparticle Dynamics in Superconducting Microwave Resonators Made of Thin-Film Granular Aluminum*, Physical Review Letters **121**, 117001 (2018), DOI: 10.1103/PhysRevLett.121.117001.

[47] K. Serniak, S. Diamond, M. Hays, S. Shankar, L. Frunzio, R. J. Schoelkopf, and M. H. Devoret: *Direct Dispersive Monitoring of Charge Parity in Offset-Charge-Sensitive Transmons*, arXiv:1903.00113 [cond-mat.mes-hall], (2019), [Online]. Available: https://arxiv.org/abs/1903.00113.

[48] G. Catelani, S. E. Nigg, S. M. Girvin, R. J. Schoelkopf, and L. I. Glazman: *Decoherence of superconducting qubits caused by quasiparticle tunneling*, Phys. Rev. B **86**, 184514 (2012), DOI: 10.1103/PhysRevB.86.184514.

[49] A. Blais, R.-S. Huang, A. Wallraff, S. M. Girvin, and R. J. Schoelkopf: *Cavity quantum electrodynamics for superconducting electrical circuits: An architecture for quantum computation*, Phys. Rev. A **69**, (2004), DOI: 10.1103/PhysRevA.69.062320.

[50] A. Wallraff, D. I. Schuster, A. Blais, L. Frunzio, R.-. S. Huang, J. Majer, S. Kumar, S. M. Girvin, and R. J. Schoelkopf: *Strong coupling of a single photon to a superconducting qubit using circuit quantum electrodynamics*, Nature **431**, 162–167 (2004), DOI: 10.1038/nature02851.

[51] M. Spiecker, *Fabrication and design of a fluxonium qubit with a granular aluminum superinductance*, Master thesis, Karlsruhe Institute of Technology (KIT), Karlsruhe, 2018.

[52] M. T. Bell, I. A. Sadovskyy, L. B. Ioffe, A. Y. Kitaev, and M. E. Gershenson: *Quantum superinductor with tunable nonlinearity*, Physical Review Letters **109**, (2012), DOI: 10.1103/PhysRevLett.109.137003.

[53] V. E. Manucharyan, *Superinductance*, Dissertation, Yale University, New Haven, CT, 2012, 162 pp.

[54] T. Weißl, G. Rastelli, I. Matei, I. M. Pop, O. Buisson, F. W. J. Hekking, and W. Guichard: *Bloch band dynamics of a Josephson junction in an inductive environment*, Phys. Rev. B **91**, 14507 (2015), DOI: 10.1103/PhysRevB.91.014507.

[55] J. Puertas Martinez, S. Leger, N. Gheeraert, R. Dassonneville, L. Planat, F. Foroughi, Y. Krupko, O. Buisson, C. Naud, W. Guichard, S. Florens, I. Snyman, and N. Roch: *A tunable Josephson platform to explore many-body quantum optics in circuit-QED*, arXiv:1802.00633 [cond-mat.mes-hall], (2018), [Online]. Available: https://arxiv.org/abs/1802.00633.

[56] T. Weißl, B. Küng, E. Dumur, A. K. Feofanov, I. Matei, C. Naud, O. Buisson, F. W. J. Hekking, and W. Guichard: *Kerr coefficients of plasma resonances in josephson junction chains*, Physical Review B **92**, (2015), DOI: 10.1103/PhysRevB.92.104508.

[57] Y. Krupko, V. D. Nguyen, T. Weißl, É. Dumur, J. Puertas, R. Dassonneville, C. Naud, F. W. J. Hekking, D. M. Basko, O. Buisson, N. Roch, and W. Hasch-Guichard: *Kerr nonlinearity in a superconducting josephson metamaterial*, Physical Review B **98**, (2018), DOI: 10.1103/PhysRevB.98.094516.

[58] C. Eichler and A. Wallraff: *Controlling the dynamic range of a josephson parametric amplifier*, EPJ Quantum Technology **1**, (2014), DOI: 10.1140/epjqt2.

[59] J. Niemeyer: *Eine einfache Methode zur Herstellung kleiner Josephson-Elemente*, PTB-Mitteilungen **84**, 251 (1974).

[60] G. J. Dolan: *Offset Masks for Lift-off Photoprocessing*, Appl. Phys. Lett. **31**, 337–339 (1977), DOI: 10.1063/1.89690.

[61] F. Lecocq, I. M. Pop, Z. Peng, I. Matei, T. Crozes, T. Fournier, C. Naud, W. Guichard, and O. Buisson: *Junction fabrication by shadow evaporation without a suspended bridge*, Nanotechnology **22**, 315302 (2011), DOI: 10.1088/0957-4484/22/31/315302.

[62] A. Kleinsasser, R. Miller, and W. Mallison: *Dependence of critical current density on oxygen exposure in Nb-AlO/sub x/-Nb tunnel junctions*, IEEE Transactions on Appiled Superconductivity **5**, 26–30 (1995), DOI: 10.1109/77.384565.

[63] L. Kuzmin and D. Haviland: *Observation of the bloch oscillations in an ultrasmall josephson junction*, Physical Review Letters **67**, 2890–2893 (1991), DOI: 10.1103/PhysRevLett.67.2890.

[64] F. Deppe, S. Saito, H. Tanaka, and H. Takayanagi: *Determination of the capacitance of nm scale josephson junctions*, Journal of Applied Physics **95**, 2607–2613 (2004), DOI: 10.1063/1.1645673.

[65] M. A. Sillanpää, T. Lehtinen, A. Paila, Y. Makhlin, L. Roschier, and P. J. Hakonen: *Direct observation of josephson capacitance*, Physical Review Letters **95**, 206806 (2005), DOI: 10.1103/PhysRevLett.95.206806.

[66] Y.-L. Wu, H. Deng, H.-F. Yu, G.-M. Xue, Y. Tian, J. Li, Y.-F. Chen, S.-P. Zhao, and D.-N. Zheng: *Fabrication of Al/AlO $_x$ /Al Josephson junctions and superconducting quantum circuits by shadow evaporation and a dynamic oxidation process*, Chinese Physics B **22**, 060309 (2013), DOI: 10.1088/1674-1056/22/6/060309.

[67] M. Tinkham, *Introduction to superconductivity*, 2. ed, ser. Dover books on physics. Mineola, NY: Dover Publ, 2004, 454 pp.

[68] T. Weißl, *Quantum phase and charge dynamics in Josephson junction chains*, Dissertation, Université de Grenoble, Grenoble, 2014.

[69] D. Walls and G. J. Milburn, Eds., *Quantum optics*, Berlin, Heidelberg: Springer Berlin Heidelberg, 2008, DOI: 10.1007/978-3-540-28574-8.

[70] D. Niepce, J. Burnett, and J. Bylander: *High Kinetic Inductance NbN Nanowire Superinductors*, arXiv:1802.01723, (2018), [Online]. Available: https://arxiv.org/abs/1802.01723.

[71] M. R. Vissers, J. Gao, D. S. Wisbey, D. A. Hite, C. C. Tsuei, A. D. Corcoles, M. Steffen, and D. P. Pappas: *Low loss superconducting titanium nitride coplanar waveguide resonators*, Applied Physics Letters **97**, 232509 (2010), DOI: 10.1063/1.3517252.

[72] N. Samkharadze, A. Bruno, P. Scarlino, G. Zheng, D. P. DiVincenzo, L. DiCarlo, and L. M. K. Vandersypen: *High-kinetic-inductance superconducting nanowire resonators for circuit QED in a magnetic field*, Physical Review Applied **5**, (2016), DOI: 10.1103/PhysRevApplied.5.044004.

[73] H. G. Leduc, B. Bumble, P. K. Day, B. H. Eom, J. Gao, S. Golwala, B. A. Mazin, S. McHugh, A. Merrill, D. C. Moore, O. Noroozian, A. D. Turner, and J. Zmuidzinas: *Titanium nitride films for ultrasensitive microresonator detectors*, Applied Physics Letters **97**, 102509 (2010), DOI: 10.1063/1.3480420.

[74] L. J. Swenson, P. K. Day, B. H. Eom, H. G. Leduc, N. Llombart, C. M. McKenney, O. Noroozian, and J. Zmuidzinas: *Operation of a titanium nitride superconducting microresonator detector in the nonlinear regime*, Journal of Applied Physics **113**, 104501 (2013), DOI: 10.1063/1.4794808.

[75] R. Barends, N. Vercruyssen, A. Endo, P. J. de Visser, T. Zijlstra, T. M. Klapwijk, and J. J. A. Baselmans: *Reduced frequency noise in superconducting resonators*, Applied Physics Letters **97**, 033507 (2010), DOI: 10.1063/1.3467052.

[76] G. J. Grabovskij, L. J. Swenson, O. Buisson, C. Hoffmann, A. Monfardini, and J.-C. Villégier: *In-situ measurement of the permittivity of helium using microwave NbN resonators*, Applied Physics Letters **93**, 134102 (2008), DOI: 10.1063/1.2996263.

[77] J. Luomahaara, V. Vesterinen, L. Grönberg, and J. Hassel: *Kinetic inductance magnetometer*, Nature Communications **5**, 4872 (2014), DOI: 10.1038/ncomms5872.

[78] B. Ho Eom, P. K. Day, H. G. LeDuc, and J. Zmuidzinas: *A wideband, low-noise superconducting amplifier with high dynamic range*, Nature Physics **8**, 623–627 (2012), DOI: 10.1038/nphys2356.

[79] N. Maleeva, L. Grünhaupt, T. Klein, F. Levy-Bertrand, O. Dupre, M. Calvo, F. Valenti, P. Winkel, F. Friedrich, W. Wernsdorfer, A. V. Ustinov, H. Rotzinger, A. Monfardini, M. V. Fistul, and I. M. Pop: *Circuit quantum electrodynamics of granular aluminum resonators*, Nature Communications **9**, 3889 (2018), DOI: 10.1038/s41467-018-06386-9.

[80] M. Peruzzo, "Geometric Superinductors for Mesoscopic Physics", Boston (USA), 2019.

[81] J. Gao, *The physics of superconducting microwave resonators*, PhD thesis, California Institute of Technology, 2008, [Online]. Available: http://resolver.caltech.edu/CaltechETD:etd-06092008-235549.

[82] F. Levy-Bertrand, T. Klein, T. Grenet, O. Dupré, A. Benoît, A. Bideaud, O. Bourrion, M. Calvo, A. Catalano, A. Gomez, J. Goupy, L. Grünhaupt, U. v. Luepke, N. Maleeva, F. Valenti, I. M. Pop, and A. Monfardini: *Electrodynamics of granular aluminum from superconductor to insulator: Observation of collective superconducting modes*, Physical Review B **99**, 094506 (2019), DOI: 10.1103/PhysRevB.99.094506.

[83] G. Deutscher, H. Fenichel, M. Gershenson, E. Grünbaum, and Z. Ovadyahu: *Transition to zero dimensionality in granular aluminum superconducting films*, J. Low Temp. Phys. **10**, 231–243 (1973), DOI: 10.1007/BF00655256.

[84] U. S. Pracht, N. Bachar, L. Benfatto, G. Deutscher, E. Farber, M. Dressel, and M. Scheffler: *Enhanced cooper pairing versus suppressed phase coherence shaping the superconducting dome in coupled aluminum nanograins*, Phys. Rev. B **93**, (2016), DOI: 10.1103/PhysRevB.93.100503.

[85] D. Kajfez: *Linear fractional curve fitting for measurement of high Q factors*, IEEE Transactions on Microwave Theory and Techniques **42**, 1149–1153 (1994), DOI: 10.1109/22.299749.

[86] S. Shahid, J. Ball, C. Wells, and P. Wen: *Reflection type q-factor measurement using standard least squares methods*, IET Microwaves, Antennas & Propagation **5**, 426 (2011), DOI: 10.1049/iet-map.2010.0395.

[87] M. S. Khalil, M. J. A. Stoutimore, F. C. Wellstood, and K. D. Osborn: *An analysis method for asymmetric resonator transmission applied to superconducting devices*, Journal of Applied Physics **111**, 054510 (2012), DOI: 10.1063/1.3692073.

[88] S. Probst, F. B. Song, P. A. Bushev, A. V. Ustinov, and M. Weides: *Efficient and robust analysis of complex scattering data under noise in microwave resonators*, Review of Scientific Instruments **86**, 024706 (2015), DOI: 10.1063/1.4907935.

[89] J. Gao, M. Daal, A. Vayonakis, S. Kumar, J. Zmuidzinas, B. Sadoulet, B. A. Mazin, P. K. Day, and H. G. Leduc: *Experimental evidence for a surface distribution of two-level systems in superconducting lithographed microwave resonators*, Applied Physics Letters **92**, 152505 (2008), DOI: 10.1063/1.2906373.

[90] J. Wenner, R. Barends, R. C. Bialczak, Y. Chen, J. Kelly, E. Lucero, M. Mariantoni, A. Megrant, P. J. J. O'Malley, D. Sank, A. Vainsencher, H. Wang, T. C. White, Y. Yin, J. Zhao, A. N. Cleland, and J. M. Martinis: *Surface loss simulations of superconducting coplanar waveguide resonators*, Applied Physics Letters **99**, 113513 (2011), DOI: 10.1063/1.3637047.

[91] C. Wang, C. Axline, Y. Y. Gao, T. Brecht, Y. Chu, L. Frunzio, M. H. Devoret, and R. J. Schoelkopf: *Surface participation and dielectric loss in superconducting qubits*, Applied Physics Letters **107**, 162601 (2015), DOI: 10.1063/1.4934486.

[92] G. Calusine, A. Melville, W. Woods, R. Das, C. Stull, V. Bolkhovsky, D. Braje, D. Hover, D. K. Kim, X. Miloshi, D. Rosenberg, A. Sevi, J. L. Yoder, E. Dauler, and W. D. Oliver: *Analysis and mitigation of interface losses in trenched superconducting coplanar waveguide resonators*, Applied Physics Letters **112**, 062601 (2018), DOI: 10.1063/1.5006888.

[93] W. Woods, G. Calusine, A. Melville, A. Sevi, E. Golden, D. K. Kim, D. Rosenberg, J. L. Yoder, and W. D. Oliver: *Determining interface dielectric losses in superconducting coplanar waveguide resonators*, arXiv:1808.10347 [quant-ph], (2018), [Online]. Available: https://arxiv.org/abs/1808.10347.

[94] S. J. Weber, K. W. Murch, D. H. Slichter, R. Vijay, and I. Siddiqi: *Single crystal silicon capacitors with low microwave loss in the single photon regime*, Applied Physics Letters **98**, 172510 (2011), DOI: 10.1063/1.3583449.

[95] L. Grünhaupt, U. von Lüpke, D. Gusenkova, S. T. Skacel, N. Maleeva, S. Schlör, A. Bilmes, H. Rotzinger, A. V. Ustinov, M. Weides, and I. M. Pop: *An argon ion beam milling process for native AlOx layers enabling coherent superconducting contacts*, Applied Physics Letters **111**, 072601 (2017), DOI: 10.1063/1.4990491.

[96] L. J. Swenson, A. Cruciani, A. Benoit, M. Roesch, C. S. Yung, A. Bideaud, and A. Monfardini: *High-speed phonon imaging using frequency-multiplexed kinetic inductance detectors*, Applied Physics Letters **96**, 263511 (2010), DOI: 10.1063/1.3459142.

[97] D. C. Moore, S. R. Golwala, B. Bumble, B. Cornell, P. K. Day, H. G. LeDuc, and J. Zmuidzinas: *Position and energy-resolved particle detection using phonon-mediated microwave kinetic inductance detectors*, Applied Physics Letters **100**, 232601 (2012), DOI: 10.1063/1.4726279.

[98] L. Cardani, I. Colantoni, A. Cruciani, S. Di Domizio, M. Vignati, F. Bellini, N. Casali, M. G. Castellano, A. Coppolecchia, C. Cosmelli, and C. Tomei: *Energy resolution and efficiency of phonon-mediated kinetic inductance detectors for light detection*, Applied Physics Letters **107**, 093508 (2015), DOI: 10.1063/1.4929977.

[99] P. K. Day, H. G. LeDuc, B. A. Mazin, A. Vayonakis, and J. Zmuidzinas: *A broadband superconducting detector suitable for use in large arrays*, Nature **425**, 817–821 (2003), DOI: 10.1038/nature02037.

[100] A. Monfardini, L. J. Swenson, *et al.*: *NIKA: A millimeter-wave kinetic inductance camera*, Astronomy and Astrophysics **521**, A29 (2010), DOI: 10.1051/0004-6361/201014727.

[101] U. Vool, I. M. Pop, K. Sliwa, B. Abdo, C. Wang, T. Brecht, Y. Y. Gao, S. Shankar, M. Hatridge, G. Catelani, M. Mirrahimi, L. Frunzio, R. J. Schoelkopf, L. I. Glazman, and M. H. Devoret: *Non-poissonian quantum jumps of a fluxonium qubit due to quasiparticle excitations*, Phys. Rev. Lett. **113**, (2014), DOI: 10.1103/PhysRevLett.113.247001.

[102] A. Kou, W. C. Smith, U. Vool, I. M. Pop, K. M. Sliwa, M. H. Hatridge, L. Frunzio, and M. H. Devoret: *Simultaneous monitoring of fluxonium qubits in a waveguide*, Phys. Rev. Appl. **9**, 064022 (2017), DOI: https://doi.org/10.1103/PhysRevApplied.9.064022.

[103] N. F. Ramsey: *A molecular beam resonance method with separated oscillating fields*, Phys. Rev. **78**, 695–699 (1950), DOI: 10.1103/PhysRev.78.695.

[104] S. Gustavsson, F. Yan, G. Catelani, J. Bylander, A. Kamal, J. Birenbaum, D. Hover, D. Rosenberg, G. Samach, A. P. Sears, S. J. Weber, J. L. Yoder, J. Clarke, A. J. Kerman, F. Yoshihara, Y. Nakamura, T. P. Orlando, and W. D. Oliver: *Suppressing relaxation in superconducting qubits by quasiparticle pumping*, Science **354**, 1573–1577 (2016), DOI: 10.1126/science.aah5844.

[105] Y.-H. Lin, L. B. Nguyen, N. Grabon, J. San Miguel, N. Pankratova, and V. E. Manucharyan: *Demonstration of protection of a superconducting qubit from energy decay*, Phys. Rev. Lett. **120**, (2018), DOI: 10.1103/PhysRevLett.120.150503.

[106] N. Bachar, *Spin-flip Scattering in Superconducting Granular Aluminum Films*, PhD thesis, Tel Aviv University, Tel Aviv (Israel), 2014.

[107] N. Bachar, S. Lerer, S. Hacohen-Gourgy, B. Almog, and G. Deutscher: *Kondo-like behavior near the metal-to-insulator transition of nanoscale granular aluminum*, Physical Review B **87**, (2013), DOI: 10.1103/PhysRevB.87.214512.

[108] S. Lerer, N. Bachar, G. Deutscher, and Y. Dagan: *Nernst effect beyond the coherence critical field of a nanoscale granular superconductor*, Physical Review B **90**, (2014), DOI: 10.1103/PhysRevB.90.214521.

[109] A. Moshe, N. Bachar, S. Lerer, Y. Lereah, and G. Deutscher: *Multi-level kondo effect and enhanced critical temperature in nanoscale granular al*, Journal of Superconductivity and Novel Magnetism **31**, 733–736 (2018), DOI: 10.1007/s10948-017-4330-z.

[110] N. Bachar, A. Levy, T. Prokscha, A. Suter, E. Morenzoni, Z. Salman, and G. Deutscher: *Kubo spins in nano-scale aluminum grains: A muon spin relaxation study*, arXiv:1803.03177, (2018), [Online]. Available: https://arxiv.org/pdf/1803.03177.pdf.

[111] A. G. Moshe, E. Farber, and G. Deutscher: *Optical conductivity of granular aluminum films near the metal to insulator transition: evidence for a BCS-BEC crossover*, arXiv:1901.02814, (2019), [Online]. Available: https://arxiv.org/pdf/1901.02814.pdf.

[113] W. Buckel and R. Hilsch: *Einfluss der Kondensation bei tiefen Temperaturen auf den elektrischen Widerstand und die Supraleitung für verschiedene Metalle*, Zeitschrift für Physik **138**, 109–120 (1954), DOI: 10.1007/BF01337903.

[114] B. Abeles, R. W. Cohen, and G. W. Cullen: *Enhancement of superconductivity in metal films*, Physical Review Letters **17**, 632–634 (1966), DOI: 10.1103/PhysRevLett.17.632.

[112] J. Bardeen, L. N. Cooper, and J. R. Schrieffer: *Theory of superconductivity*, Physical Review **108**, 1175–1204 (1957), DOI: 10.1103/PhysRev.108.1175.

[115] R. W. Cohen and B. Abeles: *Superconductivity in granular aluminum films*, Phys. Rev. **168**, 444–450 (1968), DOI: 10.1103/PhysRev.168.444.

[116] R. H. Parmenter: *Isospin formulation of the theory of a granular superconductor*, Physical Review **154**, 353–368 (1967), DOI: 10.1103/PhysRev.154.353.

[117] N. Bachar, U. S. Pracht, E. Farber, M. Dressel, G. Deutscher, and M. Scheffler: *Signatures of unconventional superconductivity in granular aluminum*, Journal of Low Temperature Physics **179**, 83–89 (2015), DOI: 10.1007/s10909-014-1244-z.

[118] U. S. Pracht, T. Cea, N. Bachar, G. Deutscher, E. Farber, M. Dressel, M. Scheffler, C. Castellani, A. M. García-García, and L. Benfatto: *Optical signatures of the superconducting goldstone mode in granular aluminum: Experiments and theory*, Physical Review B **96**, (2017), DOI: 10.1103/PhysRevB.96.094514.

[119] D. C. Mattis and J. Bardeen: *Theory of the anomalous skin effect in normal and superconducting metals*, Physical Review **111**, 412–417 (1958), DOI: 10.1103/PhysRev.111.412.

[120] V. J. Emery and S. A. Kivelson: *Importance of phase fluctuations in superconductors with small superfluid density*, Nature **374**, 434–437 (1995), DOI: 10.1038/374434a0.

[121] G. Deutscher, M. Gershenson, E. Grünbaum, and Y. Imry: *Granular superconducting films*, Journal of Vacuum Science and Technology **10**, 697–701 (1973), DOI: 10.1116/1.1318416.

[122] N. Bachar, S. Lerer, A. Levy, S. Hacohen-Gourgy, B. Almog, H. Saadaoui, Z. Salman, E. Morenzoni, and G. Deutscher: *Mott transition in granular aluminum*, Physical Review B **91**, (2015), DOI: 10.1103/PhysRevB.91.041123.

[123] C. Thompson and J. Blatt: *Shape resonances in superconductors - II simplified theory*, Physics Letters **5**, 6–9 (1963), DOI: 10.1016/S0375-9601(63)80003-1.

[124] R. H. Parmenter: *Size effect in a granular superconductor*, Physical Review **166**, 392–396 (1968), DOI: 10.1103/PhysRev.166.392.

[125] V. Ginzburg: *On surface superconductivity*, Physics Letters **13**, 101–102 (1964), DOI: 10.1016/0031-9163(64)90672-9.

[126] P. G. De Gennes: *Boundary effects in superconductors*, Reviews of Modern Physics **36**, 225–237 (1964), DOI: 10.1103/RevModPhys.36.225.

[127] O. Kammerer and M. Strongin: *Superconductivity in tungsten films*, Physics Letters **17**, 224–225 (1965), DOI: 10.1016/0031-9163(65)90496-8.

[128] M. D. Croitoru, A. A. Shanenko, and F. M. Peeters: *Dependence of superconducting properties on the size and shape of a nanoscale superconductor: From nanowire to film*, Physical Review B **76**, 024511 (2007), DOI: 10.1103/PhysRevB.76.024511.

[129] V. Z. Kresin and Y. N. Ovchinnikov: *Shell structure and strengthening of superconducting pair correlation in nanoclusters*, Physical Review B **74**, 024514 (2006), DOI: 10.1103/PhysRevB.74.024514.

[130] A. M. García-García, J. D. Urbina, E. A. Yuzbashyan, K. Richter, and B. L. Altshuler: *BCS superconductivity in metallic nanograins: Finite-size corrections, low-energy excitations, and robustness of shell effects*, Physical Review B **83**, 014510 (2011), DOI: 10.1103/PhysRevB.83.014510.

[131] Z. Lindenfeld, E. Eisenberg, and R. Lifshitz: *Possibility of electron pairing in small metallic nanoparticles*, Physical Review B **84**, 064532 (2011), DOI: 10.1103/PhysRevB.84.064532.

[132] J. Mayoh and A. M. García-García: *Strong enhancement of bulk superconductivity by engineered nanogranularity*, Physical Review B **90**, 134513 (2014), DOI: 10.1103/PhysRevB.90.134513.

[133] B. Abeles: *Effect of charging energy on superconductivity in granular metal films*, Physical Review B **15**, 2828–2829 (1977), DOI: 10.1103/PhysRevB.15.2828.

[134] B. Abeles and J. Hanak: *Superconducting and semiconducting phases of granular films*, Physics Letters A **34**, 165–166 (1971), DOI: 10.1016/0375-9601(71)90808-5.

[135] G. Deutscher: *Critical fluctuations in granular superconductors*, Physics Letters A **35**, 28–29 (1971), DOI: 10.1016/0375-9601(71)90017-X.

[136] G. Deutscher, Y. Imry, and L. Gunther: *Superconducting phase transitions in granular systems*, Physical Review B **10**, 4598–4606 (1974), DOI: 10.1103/PhysRevB.10.4598.

[137] C. Rigetti, J. M. Gambetta, S. Poletto, B. L. T. Plourde, J. M. Chow, A. D. Córcoles, J. A. Smolin, S. T. Merkel, J. R. Rozen, G. A. Keefe, M. B. Rothwell, M. B. Ketchen, and M. Steffen: *Superconducting qubit in a waveguide cavity with a coherence time approaching 0.1 ms*, Physical Review B **86**, (2012), DOI: 10.1103/PhysRevB.86.100506.

[138] M. Reagor, W. Pfaff, C. Axline, R. W. Heeres, N. Ofek, K. Sliwa, E. Holland, C. Wang, J. Blumoff, K. Chou, M. J. Hatridge, L. Frunzio, M. H. Devoret, L. Jiang, and R. J. Schoelkopf: *Quantum memory with millisecond coherence in circuit QED*, Physical Review B **94**, (2016), DOI: 10.1103/PhysRevB.94.014506.

[139] A. Romanenko, R. Pilipenko, S. Zorzetti, D. Frolov, M. Awida, S. Posen, and A. Grassellino: *Three-dimensional superconducting resonators at T$\leq$20 mK with the photon lifetime up to τ=2 seconds*, arXiv:1810.03703 [quant-ph], (2018), [Online]. Available: https://arxiv.org/abs/1810.03703.

[140] H. Rotzinger, S. T. Skacel, M. Pfirrmann, J. N. Voss, J. Münzberg, S. Probst, P. Bushev, M. P. Weides, A. V. Ustinov, and J. E. Mooij: *Aluminium-oxide wires for superconducting high kinetic inductance circuits*, Supercond. Sci. Technol. **30**, 025002 (2017), DOI: 10.1088/0953-2048/30/2/025002.

[141] F. Valenti, F. Henriques, G. Catelani, N. Maleeva, L. Grünhaupt, U. von Lüpke, S. T. Skacel, P. Winkel, A. Bilmes, A. V. Ustinov, J. Goupy, M. Calvo, A. Benoit, F. Levy-Bertrand, A. Monfardini, and I. M. Pop: *Interplay between kinetic inductance, nonlinearity, and quasiparticle dynamics in granular aluminum microwave kinetic inductance detectors*, Physical Review Applied **11**, 054087 (2019), DOI: 10.1103/PhysRevApplied.11.054087.

[142] D. Hilbert: *Über die stetige Abbildung einer Line auf ein Flächenstück*, Mathematische Annalen **38**, 459–460 (1891), DOI: 10.1007/BF01199431.

[143] C. Müller, J. H. Cole, and J. Lisenfeld: *Towards understanding two-level-systems in amorphous solids - Insights from quantum circuits*, arXiv:1705.01108 [cond-mat.mes-hall], (2017), [Online]. Available: https://arxiv.org/abs/1705.01108.

[144] S. Hunklinger, W. Arnold, S. Stein, R. Nava, and K. Dransfeld: *Saturation of the ultrasonic absorption in vitreous silica at low temperatures*, Physics Letters A **42**, 253–255 (1972), DOI: 10.1016/0375-9601(72)90884-5.

[145] B. Golding, J. E. Graebner, B. I. Halperin, and R. J. Schutz: *Nonlinear phonon propagation in fused silica below 1 k*, Physical Review Letters **30**, 223–226 (1973), DOI: 10.1103/PhysRevLett.30.223.

[146] A. D'Addabbo, *Applications of Kinetic Inductance Detectors to Astronomy and Particle Physics*, Dissertation, Université de Grenoble, 2014, [Online]. Available: https://tel.archives-ouvertes.fr/tel-01368000.

[147] K. Karatsu, A. Endo, J. Bueno, P. J. de Visser, R. Barends, D. J. Thoen, V. Murugesan, N. Tomita, and J. J. A. Baselmans: *Mitigation of cosmic ray effect on microwave kinetic inductance detector arrays*, Applied Physics Letters **114**, 032601 (2019), DOI: 10.1063/1.5052419.

[148] I. Nsanzineza and B. L. T. Plourde: *Trapping a single vortex and reducing quasiparticles in a superconducting resonator*, Physical Review Letters **113**, (2014), DOI: 10.1103/PhysRevLett.113.117002.

[149] R. Barends, J. Wenner, M. Lenander, Y. Chen, R. C. Bialczak, J. Kelly, E. Lucero, P. O'Malley, M. Mariantoni, D. Sank, H. Wang, T. C. White, Y. Yin, J. Zhao, A. N. Cleland, J. M. Martinis, and J. J. A. Baselmans: *Minimizing quasiparticle generation from stray infrared light in superconducting quantum circuits*, Applied Physics Letters **99**, 113507 (2011), DOI: 10.1063/1.3638063.

[150] A. Bespalov, M. Houzet, J. S. Meyer, and Y. V. Nazarov: *Theoretical model to explain excess of quasiparticles in superconductors*, Physical Review Letters **117**, (2016), DOI: 10.1103/PhysRevLett.117.117002.

[151] A. Rothwarf and B. N. Taylor: *Measurement of recombination lifetimes in superconductors*, Physical Review Letters **19**, 27–30 (1967), DOI: 10.1103/PhysRevLett.19.27.

[152] S. Richer, N. Maleeva, S. T. Skacel, I. M. Pop, and D. DiVincenzo: *Inductively shunted transmon qubit with tunable transverse and longitudinal coupling*, Physical Review B **96**, (2017), DOI: 10.1103/PhysRevB.96.174520.

[153] P. Winkel: *Dimer Josephson junction array amplifier*, in preparation, (2019).

[154] A. A. Houck, J. A. Schreier, B. R. Johnson, J. M. Chow, J. Koch, J. M. Gambetta, D. I. Schuster, L. Frunzio, M. H. Devoret, S. M. Girvin, and R. J. Schoelkopf: *Controlling the spontaneous emission of a superconducting transmon qubit*, Physical Review Letters **101**, (2008), DOI: 10.1103/PhysRevLett.101.080502.

[155] X. Wu, J. L. Long, H. S. Ku, R. E. Lake, M. Bal, and D. P. Pappas: *Overlap junctions for high coherence superconducting qubits*, arXiv:1705.08993 [cond-mat.supr-con], (2017), [Online]. Available: https://arxiv.org/abs/1705.08993.

[156] A. Dunsworth, A. Megrant, *et al.*: *Characterization and reduction of capacitive loss induced by sub-micron josephson junction fabrication in superconducting qubits*, Applied Physics Letters **111**, 022601 (2017), DOI: 10.1063/1.4993577.

[157] L. Sun, L. DiCarlo, M. D. Reed, G. Catelani, L. S. Bishop, D. I. Schuster, B. R. Johnson, G. A. Yang, L. Frunzio, L. Glazman, M. H. Devoret, and R. J. Schoelkopf: *Measurements of quasiparticle tunneling dynamics in a band-gap-engineered transmon qubit*, Physical Review Letters **108**, (2012), DOI: 10.1103/PhysRevLett.108.230509.

[158] H. R. Kaufman: *Technology of ion beam sources used in sputtering*, Journal of Vacuum Science and Technology **15**, 272 (1978), DOI: 10.1116/1.569569.

[159] I. M. Pop, T. Fournier, T. Crozes, F. Lecocq, I. Matei, B. Pannetier, O. Buisson, and W. Guichard: *Fabrication of stable and reproducible submicron tunnel junctions*, J. Vac. Sci. Technol. B **30**, 010607 (2012), DOI: 10.1116/1.3673790.

[160] P. Kumar, S. Sendelbach, M. A. Beck, J. W. Freeland, Z. Wang, H. Wang, C. C. Yu, R. Q. Wu, D. P. Pappas, and R. McDermott: *Origin and Reduction of $1/f$ Magnetic Flux Noise in Superconducting Devices*, Phys. Rev. Applied **6**, 41001 (2016), DOI: 10.1103/PhysRevApplied.6.041001.

[161] A. Kou, W. C. Smith, U. Vool, R. T. Brierley, H. Meier, L. Frunzio, S. M. Girvin, L. I. Glazman, and M. H. Devoret: *Fluxonium-based artificial molecule with a tunable magnetic moment*, Physical Review X **7**, (2017), DOI: 10.1103/PhysRevX.7.031037.

[162] H. l. Sueur, A. Svilans, N. Bourlet, A. Murani, L. Bergé, L. Dumoulin, and P. Joyez: *Microscopic charged fluctuators as a limit to the coherence of disordered superconductor devices*, arXiv:1810.12801 [cond-mat], (2018), [Online]. Available: http://arxiv.org/abs/1810.12801 (visited on 05/05/2019).

[163] P. J. Koppinen, L. M. Väistö, and I. J. Maasilta: *Complete stabilization and improvement of the characteristics of tunnel junctions by thermal annealing*, Applied Physics Letters **90**, 053503 (2007), DOI: 10.1063/1.2437662.

[164] M. Stepanova and S. Dew, Eds., *Nanofabrication: techniques and principles*, Wien ; New York: Springer, 2012, 344 pp.

[165] L. E. Ocola and A. Stein: *Effect of cold development on improvement in electron-beam nanopatterning resolution and line roughness*, Journal of Vacuum Science & Technology B: Microelectronics and Nanometer Structures **24**, 3061 (2006), DOI: 10.1116/1.2366698.

[166] S. Yasin, D. Hasko, and H. Ahmed: *Comparison of MIBK/IPA and water/IPA as PMMA developers for electron beam nanolithography*, Microelectronic Engineering **61-62**, 745–753 (2002), DOI: 10.1016/S0167-9317(02)00468-9.

[167] S. B. Kaplan, C. C. Chi, D. N. Langenberg, J. J. Chang, S. Jafarey, and D. J. Scalapino: *Quasiparticle and phonon lifetimes in superconductors*, Physical Review B **14**, 4854–4873 (1976), DOI: 10.1103/PhysRevB.14.4854.

List of publications

1. J. Braumüller, M. Sandberg, M. R. Vissers, A. Schneider, S. Schlör, **L. Grünhaupt**, H. Rotzinger, M. Marthaler, A. Lukashenko, A. Dieter, A. V. Ustinov, M. Weides, and D. P. Pappas: *Concentric transmon qubit featuring fast tunability and an anisotropic magnetic dipole moment*, Applied Physics Letters **108**, 032601 (2016), DOI: 10.1063/1.4940230.

2. **L. Grünhaupt**, U. von Lüpke, D. Gusenkova, S. T. Skacel, N. Maleeva, S. Schlör, A. Bilmes, H. Rotzinger, A. V. Ustinov, M. Weides, and I. M. Pop: *An argon ion beam milling process for native AlOx layers enabling coherent superconducting contacts*, Applied Physics Letters **111**, 072601 (2017), DOI: 10.1063/1.4990491.

3. P. R. Muppalla, O. Gargiulo, S. I. Mirzaei, B. P. Venkatesh, M. L. Juan, **L. Grünhaupt**, I. M. Pop, and G. Kirchmair: *Bistability in a mesoscopic josephson junction array resonator*, Physical Review B **97**, 024518 (2018), DOI: 10.1103/PhysRevB.97.024518.

4. **L. Grünhaupt**, N. Maleeva, S. T. Skacel, M. Calvo, F. Levy-Betrand, A. V. Ustinov, H. Rotzinger, A. Monfardini, G. Catelani, and I. M. Pop: *Loss Mechanisms and Quasiparticle Dynamics in Superconducting Microwave Resonators Made of Thin-Film Granular Aluminum*, Physical Review Letters **121**, 117001 (2018), DOI: 10.1103/PhysRevLett.121.117001.

5. N. Maleeva, **L. Grünhaupt**, T. Klein, F. Levy-Bertrand, O. Dupre, M. Calvo, F. Valenti, P. Winkel, F. Friedrich, W. Wernsdorfer, A. V. Ustinov, H. Rotzinger, A. Monfardini, M. V. Fistul, and I. M. Pop: *Circuit quantum electrodynamics of granular aluminum resonators*, Nature Communications **9**, 3889 (2018), DOI: 10.1038/s41467-018-06386-9.

6. F. Levy-Bertrand, T. Klein, T. Grenet, O. Dupré, A. Benoît, A. Bideaud, O. Bourrion, M. Calvo, A. Catalano, A. Gomez, J. Goupy, **L. Grünhaupt**, U. v. Luepke, N. Maleeva, F. Valenti, I. M. Pop, and A. Monfardini: *Electrodynamics of granular aluminum from superconductor to insulator: Observation of collective superconducting modes*, Physical Review B **99**, 094506 (2019), DOI: 10.1103/PhysRevB.99.094506.

7. **L. Grünhaupt**, M. Spiecker, D. Gusenkova, N. Maleeva, S. T. Skacel, I. Takmakov, F. Valenti, P. Winkel, H. Rotzinger, W. Wernsdorfer, A. V. Ustinov, and I. M. Pop: *Granular aluminium as a superconducting material for high-impedance quantum circuits*, Nature Materials **18**, 816–819 (2019), DOI: 10.1038/s41563-019-0350-3.

8. F. Valenti, F. Henriques, G. Catelani, N. Maleeva, **L. Grünhaupt**, U. von Lüpke, S. T. Skacel, P. Winkel, A. Bilmes, A. V. Ustinov, J. Goupy, M. Calvo, A. Benoit, F. Levy-Bertrand, A. Monfardini, and I. M. Pop: *Interplay between kinetic inductance, nonlinearity, and quasiparticle dynamics in granular aluminum microwave kinetic inductance detectors*, Physical Review Applied **11**, 054087 (2019), DOI: 10.1103/PhysRevApplied.11.054087.

Experimental Condensed Matter Physics
(ISSN 2191-9925)

Herausgeber
Physikalisches Institut

Prof. Dr. David Hunger
Prof. Dr. Alexey Ustinov
Prof. Dr. Georg Weiß
Prof. Dr. Wolfgang Wernsdorfer
Prof. Dr. Wulf Wulfhekel

Band 1 Alexey Feofanov
Experiments on flux qubits with pi-shifters. 2011
ISBN 978-3-86644-644-1

Band 2 Stefan Schmaus
Spintronics with individual metal-organic molecules. 2011
ISBN 978-3-86644-649-6

Band 3 Marc Müller
Elektrischer Leitwert von magnetostriktiven Dy-Nanokontakten. 2011
ISBN 978-3-86644-726-4

Band 4 Torben Peichl
Einfluss mechanischer Deformation auf atomare Tunnelsysteme – untersucht mit Josephson Phasen-Qubits. 2012
ISBN 978-3-86644-837-7

Band 5 Dominik Stöffler
Herstellung dünner metallischer Brücken durch Elektromigration und Charakterisierung mit Rastersondentechniken. 2012
ISBN 978-3-86644-843-8

Band 6 Tihomir Tomanic
Untersuchung des elektronischen Oberflächenzustands von Ag-Inseln auf supraleitendem Niob (110). 2012
ISBN 978-3-86644-898-8

Band 7 Lukas Gerhard
Magnetoelectric coupling at metal surfaces. 2013
ISBN 978-3-7315-0063-6

Band 8 Nicht erschienen.

Die Bände sind unter www.ksp.kit.edu als PDF frei verfügbar oder als Druckausgabe bestellbar.

Band 9	Jochen Zimmer **Cooper pair transport in arrays of Josephson junctions.** 2014 ISBN 978-3-7315-0130-5
Band 10	Oliver Berg **Elektrischer Transport durch Nanokontakte von Selten-Erd-Metallen.** 2014 ISBN 978-3-7315-0209-8
Band 11	Grigorij Jur'evic Grabovskij **Investigation of coherent microscopic defects inside the tunneling barrier of a Josephson junction.** 2014 ISBN 978-3-7315-0210-4
Band 12	Cornelius Thiele **STM Characterization of Phenylene-Ethynylene Oligomers on Au(111) and their Integration into Carbon Nanotube Nanogaps.** 2014 ISBN 978-3-7315-0235-7
Band 13	Michael Peter Schackert **Scanning Tunneling Spectroscopy on Electron-Boson Interactions in Superconductors.** 2014 ISBN 978-3-7315-0238-8
Band 14	Susanne Butz **One-Dimensional Tunable Josephson Metamaterials.** 2014 ISBN 978-3-7315-0271-5
Band 15	Philipp Jung **Nonlinear Effects in Superconducting Quantum Interference Meta-Atoms.** 2014 ISBN 978-3-7315-0294-4
Band 16	Sebastian Probst **Hybrid quantum system based on rare earth doped crystals.** 2016 ISBN 978-3-7315-0345-3
Band 17	Wolfram Kittler **Magnetische Anisotropie und Quantenphasenübergang in $CeTi_{1-x}V_xGe_3$.** 2015 ISBN 978-3-7315-0363-7
Band 18	Moritz Peter **Towards magnetic resonance in scanning tunneling microscopy using heterodyne detection.** 2015 ISBN 978-3-7315-0410-8

Band 19 Junji Tobias Märkl
Investigation of Magnetic Adatoms with Scanning Tunneling Techniques. 2015
ISBN 978-3-7315-0435-1

Band 20 Jochen Braumüller
Quantum simulation experiments with superconducting circuits. 2018
ISBN 978-3-7315-0780-2

Band 21 Jinjie Chen
Local Investigation of Single Magnetic Molecules with Scanning Tunneling Microscopy. 2018
ISBN 978-3-7315-0819-9

Band 22 Arnold Seiler
Einfluss der Leitungselektronen auf die Dynamik atomarer Tunnelsysteme in ungeordneten Festkörpern: Relaxationsprozesse in metallischen Gläsern und ungeordneten dünnen Aluminiumoxid-Schichten. 2019
ISBN 978-3-7315-0870-0

Band 23 Jasmin Maria Jandke
Elastic and Inelastic Scanning Tunneling Spectroscopy on Iron-Based Superconductors. 2019
ISBN 978-3-7315-0747-5

Band 24 Kevin Edelmann
Electroluminescence from Plasmonic Excitations in a Scanning Tunnelling Microscope. 2019
ISBN 978-3-7315-0923-3

Band 25 Lei Zhang
Sub-Kelvin scanning tunneling microscopy on magnetic molecules. 2019
ISBN 978-3-86644-950-3

Band 26 Lukas Grünhaupt
Granular aluminium superinductors. 2019
ISBN 978-3-7315-0956-1